Lecture Notes in Computer Science   13863

Founding Editors

Gerhard Goos
Juris Hartmanis

Editorial Board Members

The series Lecture Notes in Computer Science (LNCS), including its subseries Lecture Notes in Artificial Intelligence (LNAI) and Lecture Notes in Bioinformatics (LNBI), has established itself as a medium for the publication of new developments in computer science and information technology research, teaching, and education.

LNCS enjoys close cooperation with the computer science R & D community, the series counts many renowned academics among its volume editors and paper authors, and collaborates with prestigious societies. Its mission is to serve this international community by providing an invaluable service, mainly focused on the publication of conference and workshop proceedings and postproceedings. LNCS commenced publication in 1973.

Andrei Sleptchenko · Angelo Sifaleras ·
Pierre Hansen

Editors

Variable Neighborhood Search

9th International Conference, ICVNS 2022
Abu Dhabi, United Arab Emirates, October 25–28, 2022
Revised Selected Papers

Springer

Editors
Andrei Sleptchenko 🆔
Khalifa University
Abu Dhabi, United Arab Emirates

Angelo Sifaleras 🆔
University of Macedonia
Thessaloniki, Greece

Pierre Hansen
HEC Montréal
Montréal, QC, Canada

ISSN 0302-9743   ISSN 1611-3349 (electronic)
Lecture Notes in Computer Science
ISBN 978-3-031-34499-2   ISBN 978-3-031-34500-5 (eBook)
https://doi.org/10.1007/978-3-031-34500-5

This Springer imprint is published by the registered company Springer Nature Switzerland AG
The registered company address is: Gewerbestrasse 11, 6330 Cham, Switzerland

Preface

This volume edited by Andrei Sleptchenko, Angelo Sifaleras, and Pierre Hansen contains peer-reviewed papers from the 9th International Conference on Variable Neighborhood Search (ICVNS 2022) held in Abu Dhabi, United Arab Emirates (UAE), during October 25–28, 2022.

The conference follows previous successful meetings that were held in Puerto de la Cruz, Tenerife, Spain (2005); Herceg Novi, Montenegro (2012); Djerba, Tunisia (2014); Malaga, Spain (2016); Ouro Preto, Brazil, (2017); Sithonia, Halkidiki, Greece (2018); Rabat, Morocco (2019); and Abu Dhabi, UAE (2021 online).

This edition was initiated by Nenad Mladenović and Andrei Sleptchenko from Khalifa University (UAE), together with Angelo Sifaleras, from the University of Macedonia (Greece). Unfortunately, Prof. Nenad Mladenović passed away on Saturday 7 of May 2022 after being hospitalized at the Cleveland Clinic in Abu Dhabi. Prof. Nenad Mladenović and Prof. Pierre Hansen were the founders of the Variable Neighborhood Search (VNS) metaheuristic in 1997. Nowadays the series of International Conferences on Variable Neighborhood Search (ICVNS) is dedicated to the VNS method and organized on a regular basis. Members of both the program and the organizing committee express their sincere condolences to the family of Prof. Nenad Mladenović.

Like its predecessors, the main goal of ICVNS 2022 was to provide a stimulating environment in which researchers coming from various scientific fields could share and discuss their knowledge, expertise and ideas related to the VNS Metaheuristic and its applications. Due to post-pandemic restrictions, ICVNS 2022 was organized in hybrid (online and offline) mode with the help of the Office of Marketing and Communications of Khalifa University.

The following three plenary lecturers shared their current research directions with the ICVNS 2022 participants:

- Panos M. Pardalos, from the Center for Applied Optimization, Department of Industrial and Systems Engineering, of the University of Florida, USA,
- Said Salhi, from the Department of Operational Research/Management Science at the Kent Business School, UK,
- Angelo Sifaleras, from the Department of Applied Informatics, University of Macedonia, Greece.

Around 60 participants took part in the ICVNS 2022 conference, and a total of 24 papers were accepted for oral presentation. A total of 11 long papers were accepted for publication in this LNCS volume after thorough, single-blind, peer-reviewing (three reviews per submission) by the members of the ICVNS 2022 Program Committee. These papers describe recent advances in methods and applications of Variable Neighborhood Search.

The editors thank all the participants in the conference for their contributions and for their continuous effort to disseminate VNS, and are grateful to the reviewers for preparing excellent reports. The editors wish to acknowledge the Springer LNCS editorial staff for their support during the entire process of making this volume. Finally, we express our gratitude to the organizers and sponsors of the ICVNS 2022 meeting:

- The Research Center for Digital Supply Chains and Operations Management, Khalifa University,
- The Office of Marketing and Communications of Khalifa University,
- The Abu Dhabi Convention and Exhibition Bureau,
- The EURO Working Group on Metaheuristics (EWG EU/ME).

Their support is greatly appreciated for making ICVNS 2022 a great scientific event.

March 2023

Andrei Sleptchenko
Angelo Sifaleras
Pierre Hansen

Organization

Conference Chairs

Pierre Hansen (General Chair)	HEC Montréal, Canada
Mohammed Omar (General Co-chair)	Khalifa University, U.A.E.
Andrei Sleptchenko (Conference Chair)	Khalifa University, U.A.E.
Angelo Sifaleras (Conference Chair)	University of Macedonia, Greece

Program Committee

Abraham Duarte	Universidad Rey Juan Carlos, Spain
Ada Alvarez	Universidad Autónoma de Nuevo León, México
Adriana Gabor	Khalifa University, U.A.E.
Ali Allahverdi	Kuwait University, Kuwait
Andrei Sleptchenko	Khalifa University, U.A.E.
Angelo Sifaleras	University of Macedonia, Greece
Anton Eremeev	Omsk State University, Russia
Athanasios Migdalas	Luleå University of Technology, Sweden
Bassem Jarboui	Higher Colleges of Technology, U.A.E.
Celso Ribeiro	Universidade Federal Fluminense, Brazil
Chandra Irawan	Nottingham University Business School, China
Christos D. Tarantilis	Athens University of Economics, Greece
Cláudio Alves	University of Minho, Braga, Portugal
Daniel Aloise	Polytechnique Montréal, Canada
Dragan Urošević	Mathematical Institute SANU, Serbia
Eduardo G. Pardo	Universidad Politécnica de Madrid, Spain
Emilio Carrizosa	University of Seville, Spain
Gilles Caporossi	HEC Montréal, Canada
Günther Raidl	TU Wien, Austria
Hasan Turan	University of New South Wales, Australia
Jack Brimberg	Royal Military College, Kingston, Canada
John Beasley	Brunel University London, UK
José A. Moreno-Perez	University of La Laguna, Spain
Jun Pei	University of Hefei, China

Karl Dörner	University of Vienna, Austria
Kenneth Sörensen	University of Antwerp, Belgium
Laureano Escudero	Universidad Rey Juan Carlos, Spain
Leonidas Pitsoulis	Aristotle University of Thessaloniki, Greece
Luiz Satoru Ochi	Fluminense Federal University, Brazil
Marc Sevaux	University of Southern Brittany, France
Marcone Jamilson Freitas Souza	UFOP, Brazil
Michel Gendreau	École Polytechnique de Montréal, Canada
Panos Pardalos	University of Florida, USA
Patrick Siarry	Université Paris-Est Créteil Val de Marne, France
Pierre Hansen	HEC Montréal, Canada
Qiuhong Zhao	Beihang University, China
Raca Todosijević	Mathematical Institute SANU, Serbia
Rachid Benmansour	INSEA, Morocco
Raja Jayaraman	Khalifa University, U.A.E.
Richard Hartl	University of Vienna, Austria
Rustam Mussabayev	National Institute of Information Security Development, Kazakhstan
Saïd Hanafi	University of Valenciennes, France
Said Salhi	University of Kent, UK
Sergio Consoli	European Commission, Joint Research Centre, Belgium
Tatiana Davidović	Mathematical Institute SANU, Serbia
Teodor G. Crainic	Université du Québec à Montréal, Canada
Varvara Raskazzova	Moscow Aviation Institute, Russia
Vera Kovačević-Vujčić	University of Belgrade, Serbia
Vitor Nazário Coelho	Universidade Federal Fluminense, Brazil
Yannis Marinakis	Technical University of Crete, Greece
Yuri Kochetov	Sobolev Institute of Mathematics, Russia
Zorica Stanimirovic	University of Belgrade, Serbia

Organizing Committee

Arif Sultan Al Hammadi	Khalifa University, U.A.E.
Sayed Al Hashmi	Khalifa University, U.A.E.
Khulood Al Ali	Khalifa University, U.A.E.
Eyad Zerba	Khalifa University, U.A.E.

Contents

A Metaheuristic Approach for Solving Monitor Placement Problem 1
Alejandra Casado, Nenad Mladenović, Jesús Sánchez-Oro,
and Abraham Duarte

A VNS-Based Heuristic for the Minimum Number of Resources Under
a Perfect Schedule 14
Rachid Benmansour

BVNS for Overlapping Community Detection 27
Sergio Pérez-Peló, Jesús Sánchez-Oro, Antonio González-Pardo,
and Abraham Duarte

A Simulation-Based Variable Neighborhood Search Approach
for Optimizing Cross-Training Policies 42
Moustafa Abdelwanis, Nenad Mladenovic, and Andrei Sleptchenko

Multi-objective Variable Neighborhood Search for Improving Software
Modularity 58
Javier Yuste, Eduardo G. Pardo, and Abraham Duarte

An Effective VNS for Delivery Districting 69
Ahmed Aly, Adriana F. Gabor, and Nenad Mladenovic

BVNS for the Minimum Sitting Arrangement Problem in a Cycle 82
Marcos Robles, Sergio Cavero, and Eduardo G. Pardo

Assigning Multi-skill Configurations to Multiple Servers with a Reduced
VNS 97
Thiago Alves de Queiroz, Beatrice Bolsi, Vinícius Loti de Lima,
Manuel Iori, and Arthur Kramer

Multi-Round Influence Maximization: A Variable Neighborhood Search
Approach 112
Isaac Lozano-Osorio, Jesús Sánchez-Oro, and Abraham Duarte

A VNS Based Heuristic for a 2D Open Dimension Problem 125
Layane Rodrigues de Souza Queiroz and Thiago Alves de Queiroz

x   Contents

A Basic Variable Neighborhood Search Approach for the Bi-objective
Multi-row Equal Facility Layout Problem 137
  Nicolás R. Uribe, Alberto Herrán, and J. Manuel Colmenar

Author Index 149

A Metaheuristic Approach for Solving Monitor Placement Problem

Alejandra Casado[1]([envelope]) [ID], Nenad Mladenović[2] [ID], Jesús Sánchez-Oro[1] [ID],
and Abraham Duarte[1] [ID]

[1] Universidad Rey Juan Carlos, Tulipán s/n., 28933 Móstoles, Spain
{alejandra.casado,jesus.sanchezoro,abraham.duarte}@urjc.es
[2] Khalifa University, Zone 1, Abu Dhabi, UAE
nenad.mladenovic@ku.ac.ae

Abstract. There are several hard combinatorial optimization problems
that, in the context of communication networks, must be solved in short
computing times since they are solving real-time critical tasks. This work
is focused on the monitor placement problem, whose objective is to locate
specific devices, called monitors, in certain nodes of a network with the
aim of performing a complete network surveillance. As a consequence of
the constant evolution of networks, the problem must be solved in real
time if possible. If a solution cannot be found in the allowed computing time, then a penalty is assumed for each link of the network which
remains uncovered. A Variable Neighborhood Search algorithm is proposed for solving this problem, comparing it with a hybrid evolutionary
algorithm over a set of instances derived from real-life networks to evaluate its efficiency and efficacy.

Keywords: monitor placement problem · basic variable neighborhood
search · local search · metaheuristics

1 Introduction

Nowadays, communications are a key part of almost every task, from profes-
sional to personal issues: streaming, banking, shopping, security, etc. In their
origins, the security of those networks was not relevant. However, the appearance of cyber-attacks and the necessity of data protection has highlighted the
relevance of having secure networks. Therefore, every modern company requires
to have their communication networks totally surveyed, even in the case of those
networks which are in continuous evolution.

A. Casado, J. Sánchez-Oro and A. Duarte research was funded by "Ministerio de Ciencia, Innovación y Universidades" under grant ref. PID2021-125709OA-C22, "Comunidad de Madrid" and "Fondos Estructurales" of European Union with grant refs.
S2018/TCS-4566, Y2018/EMT-5062. N. Mladenović has been partially supported by
the Science Committee of the Ministry of Education and Science of the Republic of
Kazakhstan, Grant No. AP08856034.

A. Sleptchenko et al. (Eds.): ICVNS 2022, LNCS 13863, pp. 1–13, 2023.
https://doi.org/10.1007/978-3-031-34500-5_1

The main objective of most of the cyber-attacks to networks is to have an important economic impact in the company suffering the attack, which may eventually lead to an economic crisis. However, some services are considered as critical, since the society cannot be maintained without them: transportation, healthcare, defense, among others. A successful attack over one or more of these services can be critical, resulting in humanitarian crisis [1].

With the increase in the number of cyber-attacks, and the rising concerns about privacy on the Internet, every company and institution are continuously working on improving their security systems. The main objective is to minimize the security breaches, thus reducing the number of attacks. Even more, in the case of suffering an attack, the faster the reaction, the smaller the probabilities of success [13].

The most extended types of cyber-attacks are Denial of Service (Dos) and its distributed variant (DDoS). This is mainly because a successful DoS or DDoS usually results in completely disabling a network. Even more, if the network under attack has several services that directly depend on it, the attack normally results in a cascade failure, damaging a large number of services [5].

The main efforts of researchers and practitioners in increasing security are focused on the early detection of potential threats: loss of nodes in the network, unauthorized access to some parts of the network, malware spreading, etc. In order to protect the network from those threats, they must be detected in almost real time. Otherwise, it would be impossible to perform a counterattack to disable the effects of the threat.

The first and one of the most important phases of network protection is the surveillance of the network. A good surveillance guarantees monitoring the complete network, allowing the administrators to provide a fast and efficient reaction to any kind of potential threat. In most of the networks, the process of monitoring consists of deploying a special type of device called monitor in certain nodes of the network. These devices are able to analyze and survey all the traffic that goes through them. Then, the administrators are able to gather and analyze the information recovered from the monitors to detect potential threats, thus helping them to protect the most critical nodes.

Taking this into account, it is easy to see that a network is completely monitored if a monitor is deployed in every node of the network, since it guarantees that any communication in the network will be analyzed. However, the cost of deploying a monitor in a network is too high to afford the deployment of one of these devices in each node of the network. Even more, deploying a monitor not only has an associated cost, but also implies an overhead in the network performance derived from capturing all the communications that go through them. Then, it is interesting to survey the network with the minimum number of monitors.

This problem is named Monitor Placement Problem (MPP), and it has been widely studied in the literature from both theoretical and practical points of view. The research on MPP has discovered several variants that consider new features of the networks under surveillance. Traditional approaches consider static networks, where changes over its original design are not common. However, this is not a realistic situation, so this research is focused in those networks that are in continuous evolution. The constant changes in the network force the administrators to select the nodes to deploy a monitor in short computing times to avoid affecting the performance of the network.

This problem is a generalization of the well-known vertex covering problem, which has been proven to be $\mathcal{NP}$-complete [12]. Then, exact approaches are not able to solve instances derived from real-life scenarios, due to the inherent complexity of those networks. Even more, the MPP is still $\mathcal{NP}$-complete for approximations within a factor smaller than 1.36 [6]. As a consequence, the MPP and related problems have been mainly tackled from a heuristic point of view. The minimum vertex cover, as well as the maximum independent set, were firstly solved using evolutionary algorithms [2,8]. Then, a combination of a branch and bound procedure and several heuristics were proposed for dealing with random graphs in the context of minimum vertex cover [14,15]. After that, different approaches were presented for solving that problem, such as genetic algorithms [11], hierarchical Bayesian algorithms [21], or simulated annealing [10], among others. The generalized vertex cover was tackled by a new genetic algorithm [16], considering a weighted network in both nodes and links. More recently, a parallel design of evolutionary algorithm was presented [4], which is able to take advantage of the hardware architecture by using distributed computation.

Several approaches have considered evolutionary algorithms for solving MPP, including a population injection method [17], or a hybrid evolutionary algorithm for the dynamic variant of the problem [18]. Even more, the best approach for solving the MPP found in the literature is also a hybrid search heuristic [19], which is an evolution of the population injection method originally proposed in [17], but focusing on the diversification of the search.

This research proposes a Variable Neighborhood Search (VNS) algorithm for dealing with the MPP. A constructive method is proposed to provide a high quality starting point for the search, as well as an intelligent local search method to reach a local optimum. Additionally, an intensified shake procedure is presented to guide the search during the diversity phase of VNS. The remaining of the paper is structured as follows. Section 2 describes the MPP, Sect. 3 presents the algorithm proposed for solving the MPP, Sect. 4 shows the computational experiments performed to test the proposal, comparing it with the best method found in the literature and, finally, Sect. 5 draws some conclusions derived from this research.

2   Problem Definition

A network is modeled as a non-directed graph $G = (V, E)$ where the set of nodes is represented by V, with $|V| = n$, and the links between nodes in the network, E (with $|E| = m$) are represented by tuples (u, v) that indicate that there exists a communication between nodes u and v, with $u, v \in V$. Let us introduce a vertex cover Λ of a network, defined as a subset of vertices $\Lambda \subseteq V$ satisfying the constraint that, for every edge $(u, v) \in E$, either $u \in \Lambda$ or $v \in \Lambda$. A minimal vertex cover is a vertex cover with the minimum size. Then, the Minimum Vertex Cover Problem is formally defined as:

$$MVCP(G) = \min_{\Lambda \in \mathcal{C}} |\Lambda|$$

where $\mathcal{C}$ is the set of all existing vertex covers for the network G under consideration. Notice that a network can have several different minimum vertex covers, so the MVCP consists of finding one of those vertex covers.

Having defined the MVCP, the MPP is a variant of the MVCP with direct application in communication networks and, therefore, there are some additional constraints that must be satisfied. The first constraint refers to the computing time available to solve the problem, which is drastically reduced when solving real-life networks when comparing it with the traditional MVCP. This reduction is directly related with its practical application: as it was aforementioned, in order to defend the network from an attack, it is required to detect it as soon as possible, allowing the administrators to carry out the corresponding measures to avoid the attack. However, in some cases it is not possible to find a solution that monitors the complete network in the available computing time, and then the objective is to find the solution which maximizes the number of connections controlled in the network. Since not all the connections have the same relevance, it is important to prioritize the connections to be surveyed [3].

In the context of MPP, a network is considered securized if and only if all the communications between nodes are covered by a monitor (at least). Then, a solution is considered optimal if it monitors the complete network with the minimum impact in its performance. In other words, the MPP tries to find a solution which monitors the complete network with the minimum number of monitors deployed in it. Without loss of generality, in the variant tackled in this research a monitor can be placed in any node of the network.

The link priority is totally related to the network nature and the relevance given by the administrators: the bandwidth of the link, the traffic flow through the link, a special type of relevance related to the nodes conforming the link, etc. Several studies have been performed to choose the correct priority for every link of the network [20, 24]. These link priorities are the key when comparing MPP with MVCP. In the original MVCP, the priorities are not considered, so the model needs from an adaptation. In particular, the priorities are usually included in the model [19] by defining a priority function $p : E \rightarrow \mathbb{N}$ which assigns a certain penalty to each link not monitored in a solution. This adaptation allows algorithms to compare solutions with the same number of monitors, considering the penalty of the uncovered links.

Following the definition of the penalty, the objective function for the MPP is divided into two different parts: the number of monitors deployed in the network, and the total penalty of the links which are not surveyed. Let us define S as a solution for the MPP which contains the nodes in which a monitor is deployed (naturally, $S \subseteq V$). The objective function of the MPP is then formally defined as:

$$MPP(S) = |S| + \sum_{e \in E'} p(u, v)$$

with E' being the set of links which are not covered by any of the selected monitors, i.e., $E' = \{(u, v) \in E : u \notin S \land v \notin S\}$. The problem then seeks to find a solution S^* with the minimum objective function value. In mathematical terms,

$$S^* = \arg \min_{S \in \mathbb{S}} MPP(S)$$

where $\mathbb{S}$ models the set of all feasible solutions for the MPP. Since, in the context of MPP, any subset of nodes is a feasible solution (assuming the corresponding penalty), the solution space is conformed with all the possible subsets of nodes that can be conformed with V. Hence, the two trivial solutions where a monitor is deployed in every node, i.e., $S = V$, and the one in which no monitors are deployed, i.e., $S = \emptyset$, are also considered.

Two factors are then considering when evaluating a solution: the number of monitors deployed and the total penalty for those uncovered links. The literature shows different ways of calculating the penalty, but in this work the same approach as in the best previous method found in the literature [19] is followed. In particular, a static linear distance penalty function is considered, which will be later described in Sect. 4.

3   Algorithmic Approach

Most of the works related to the MPP found in the literature considers evolutionary algorithms as the main approach for solving the problem. This research proposes a different point of view where a trajectory-based metaheuristic is considered, instead of a population-based metaheuristic such as evolutionary algorithms. The main difference is that trajectory-based metaheuristics performs the search by maintaining a single solution which is iteratively modified trough out different phases, while population-based metaheuristics are based on maintaining a complete population of solutions with the aim of combining it during the search.

In particular, Variable Neighborhood Search (VNS) metaheuristic is considered, whose success is based on performing systematic changes of neighborhoods to improve the quality of the solutions obtained without being a resulting in a computationally demanding procedure. VNS is in continuous evolution, as it can be seen in the large variety of schemes proposed, which usually differ in how the neighborhoods are explored. We can highlight Basic VNS, which combines both deterministic and stochastic neighborhood changes, Reduced VNS, which is focused on stochastic neighborhood exploration, and Variable Neighborhood Descent, whose success relies on deterministic changes of neighborhoods. Notwithstanding, VNS researchers have proposed several new variants in the last decades: General VNS, Variable Neighborhood Decomposition Search, Variable Formulation Search, Less Is More Approach VNS, etc. [9].

The proposed algorithm for the MPP follows the Basic VNS (BVNS), which is able to balance diversification and intensification with a perturbation method and a local improvement phase, respectively. Algorithm 1 shows the pseudocode of Basic VNS.

Algorithm 1. $BVNS(S, k_{\max}, k_{step})$

1: $k \leftarrow 1$
2: **while** $k \leq k_{\max}$ **do**
3:     $S' \leftarrow Shake(S, k)$
4:     $S'' \leftarrow Improve(S')$
5:     **if** $MPP(S'') < MPP(S)$ **then**
6:         $k \leftarrow 1$
7:         $S \leftarrow S''$
8:     **else**
9:         $k \leftarrow k + k_{step}$
10:    **end if**
11: **end while**
12: **return** S

The method requires from three input parameters: the initial solution S (see Sect. 3.1 for more details), the maximum neighborhood to be explored during the search $k_{\max}$, and the value of each step of the algorithm k_{step}. The algorithm starts with the first neighborhood (step 1) and, then, it iterates until reaching the largest predefined neighborhood $k_{\max}$ (step 2–11). In each neighborhood, the solution is perturbed following the shake procedure described in Sect. 3.3 (step 3). The perturbed solution S' is then improved using the method described in Sect. 3.2 to find a local optimum in the current neighborhood (step 4). Finally, BVNS performs the neighborhood change stage. In particular, if the improved solution S'' outperforms the best solution found so far (step 5), the method restarts the search from the first neighborhood (step 6), updating the best solution found (step 7). Otherwise, the algorithm continues with the next neighborhood (step 9). BVNS ends when $k \geq k_{\max}$ returning the best solution found during the search (step 12).

Our proposal is formed by different components, all of them detailed hereinafter. However, as a summary, the algorithm starts with an initial greedy solution. A local optimum is found from that initial solution by executing the local search. After that, the whole BVNS algorithm, showed in Algorithm 1, is executed δ times, which value is defined in Sect. 4.

3.1 Initial Solution

The initial solution required by VNS can be generated either at random or with a specific constructive procedure. Recent works on VNS have shown that providing VNS a good starting point usually results in a more robust and efficient algorithm, in terms of quality and/or computing time [23]. As a consequence, this research proposes a new constructive procedure to start the search from a promising region of the search space.

A simple yet effective greedy procedure is proposed in Algorithm 2 which, starting from an empty solution, is able to iteratively deploy new monitors until all the links are covered.

Algorithm 2. *GreedyConstructive(G)*

1: $S \leftarrow \emptyset$
2: $UL \leftarrow \{(u, v) \in E : u \notin S \wedge v \notin S\}$
3: $C \leftarrow V$
4: **while** $UL \neq \emptyset$ **do**
5:     $c \leftarrow \arg\max_{u \in C} \sum_{\substack{(u,v) \in E \\ v \in C}} p(u, v)$
6:     $S \leftarrow S \cup \{c\}$
7:     $C \leftarrow C \setminus \{c\}$
8:     $UL \leftarrow UL \setminus \{(c, x) : \forall x \in N(c)\}$
9: **end while**
10: **return** S

The algorithm starts from an empty solution S (step 1). Then, the list of uncovered links UL is created with every edge in the graph which is not covered by any monitor (step 2). The candidate nodes to host a monitor are all the nodes of the instance (step 3). The method now iterates until covering all the links in the network (steps 4–9). In each iteration, the next candidate node is selected as the one which is able to minimize the penalty that affects to the solution under construction (step 5). In other words, we only consider (by summing up the corresponding penalty) edges whose both endpoints belong to C.

The selected candidate c is then added to the solution (step 6), updating the set of candidates (step 7), removing from the set of uncovered links UL all edges in which c is involved (step 8). Notice that $N(c)$ refers to the adjacent nodes to c. The method ends returning a solution S where all the links are covered (step 10).

3.2 Improvement Method

The intensification part of VNS relies in the local improvement method used for reaching a local optimum with respect to a certain neighborhood. Notice that, although some works have proposed a complex metaheuristic in this phase [7,22], leading to successful researches, the MPP requires from a fast local search method since the time constraints are usually hard. Then, in this work a fast local search is proposed, which is conformed with three main elements: the move operator used in the method, the neighborhood of solutions that can be generated through the given move operator, and the strategy selected to explore the neighborhood.

The first element that needs to be defined is the move operator, which indicates how a solution will be modified in each step. A good move operator should be designed for reducing the number of monitors deployed without increasing the penalty caused by the uncovered links. In this work, the move operator Swap is proposed, which removes a monitor already deployed, replacing it with a new one. More formally,

$$Swap(S,u,v) \leftarrow (S \setminus \{u\}) \cup \{v\}$$

Notice that this move operator will never lead to an improvement by itself, since the number of monitors will never be reduced and the penalty may eventually increase. However, the new monitor deployed will eventually make some of the other monitors unnecessary, covering the same links. Then, if the deployment of a new monitor results in, at least, one new redundant monitor, its removal will result in an improvement. This procedure, named as purge, basically consists of traversing all the deployed monitors, checking if all edges still covered after removing each monitor. If the objective function value decreases after applying this method, then an improvement has been found in the local search.

This move operator allows us to define the second key part of a local search: the neighborhood of a given solution S. In the context of MPP, the neighborhood N_{swap} is defined as all the solutions that can be reached after performing a single Swap move. In mathematical terms,

$$N_{Swap}(S) \leftarrow \{S' \leftarrow Swap(S, u, v) : \forall u \in \bar{S} \wedge v \in (\bar{V} \setminus \bar{S})\}$$

Finally, it is necessary to indicate the strategy followed to traverse the neighborhood. Two main strategies are usually considered in the literature: first improvement and best improvement. The former performs the first move that leads to an improvement, while the latter performs the best move found in the neighborhood. Notice that best improvement methods are usually slower than first improvement ones, since they require to evaluate the complete neighborhood in each iteration, while a first improvement approach stops whenever an improvement is found. Since MPP requires from fast procedures, we select first improvement as neighborhood exploration strategy in the proposed local search. In order to avoid biasing the search, the neighbor solutions are explored at random in each iteration, increasing the diversification of the search. It is worth mentioning that the purge procedure is applied over each neighbor solution to validate if an improvement is found.

3.3   Shake

The shake procedure is responsible for the diversification part of the algorithm to avoid getting trapped in local optima during the local search phase. In particular, this method perturbs the solution under exploration by randomly applying the move operator, instead of focusing on those movements that leads to an improvement. The size of the perturbation is controlled by parameter k, indicating that k random Swap moves will be performed.

When no improvement is found, the value of the parameter is increased, since more diversification is needed by the local search method to find more promising regions of the search space. On the contrary, when an improvement is found, the perturbation starts again from the minimum size, to avoid performing an extremely large perturbation of the incumbent solution which may lead the algorithm to miss high-quality solutions.

The move operator used in the perturbation phase is the same as in the local search method, Swap. However, in this case, the node to be removed is selected completely random while the new monitor deployed is selected at random among its adjacent nodes, to assure that the resulting solution has no additional penalty. It is important to remark that the perturbation phase will not consider any move that lead to an increase in the penalty.

4 Computational Results

Once the algorithmic proposal has been presented, it is necessary to validate its efficiency and efficacy, as well as to select the best value for the input parameters of the proposed algorithm. This section is designed for tuning the parameters of the Basic VNS algorithm proposed, as well as to perform a competitive testing with the best method found in the literature for the MPP.

All the algorithms have been implemented in Java 11 and the experiments have been conducted in an AMD EPYC 7282 (2.8 GHz) and 72 GB RAM. We have considered the same set of instances as the ones presented in the related literature, where the best algorithm for the MPP is introduced [19]. Unfortunately, due to hardware constraints, instances delaunay_n17, delaunay_n18, delaunay_n19 and delaunay_n20 cannot be considered, since the available hardware does not have enough memory to store them. However, the results obtained will show that, with the appropriate hardware, the method would be able to provide good results as a consequence of its scalability. Then, a total set of instances conformed by 35 networks is considered, of which 10 representative instances have been selected to configure the proposed algorithm in order to avoid overfitting.

All the experiments report the following metrics: Avg., the average objective function value obtained by each algorithm; Time (s), the computing time required by each algorithm to finish, measured in seconds; Dev (%), the average deviation with respect to the best solution found during the experiment; and # Best, the times that the algorithm reaches the best solution of the experiment.

The first experiment analyze the quality obtained changing the values of the input parameters of BVNS. The algorithm needs just two parameters: δ, which is the number of complete BVNS iterations, and k_{max}, the maximum neighborhood to be explored in each complete iteration. The values tested for the largest neighborhood are $k_{max} = \{0.1, 0.2, 0.3, 0.4, 0.5\}$ (percentage of the number of nodes in the solution), where each value represents a percentage of the number of nodes of the instance to guarantee the scalability of the proposal. We do not consider larger values of k_{max} since perturbing more than half of the solution will result in a completely different one, which is against the philosophy of the VNS framework. Regarding the number of iterations, we test $\delta = \{1, 10, 20, 30, 40\}$. The value of k_{step} is fixed to 0.05 (5% of the number of nodes in the solution).

Since modifying one of the parameters might affect to the other one, we have decided to show the results obtained in two heat maps, where the worst values are colored in red and the best values in green, interpolating the values between them using a color gradient. Table 1 shows the results obtained.

Table 1. Heat map of the computing times (left) and average deviation (right) when considering different values of δ and k_{max}.

δ \ k_{max}	0.1	0.2	0.3	0.4	0.5	δ \ k_{max}	0.1	0.2	0.3	0.4	0.5
1	0.20	0.26	0.57	1.15	2.04	1	0.36	0.34	0.32	0.12	0.12
10	0.19	0.99	3.15	6.05	12.74	10	0.36	0.17	0.07	0.11	0.06
20	0.20	1.89	5.75	11.79	22.25	20	0.36	0.09	0.03	0.11	0.06
30	0.23	2.56	8.50	17.91	33.51	30	0.33	0.09	0.03	0.10	0.06
40	0.24	3.22	10.93	27.10	45.64	40	0.33	0.09	0.03	0.10	0.03

Analyzing the heat map of computing times (left), as expected, it grows with the number of iterations and maximum neighborhood explored, being two orders of magnitude slower in some cases. If we simultaneously analyze the heat map of average deviation (right), we can clearly see that the best values are obtained when considering small values of k_{max}, specifically $k_{max} = 0.2$, and the number of iterations stagnates when reaching 20. Since the computing time of $k_{max} = 0.2$ and $\delta = 20$ is also one of the smallest in the experiment, we select these parameter values.

Having defined the best parameter values for k_{max} and δ, it is necessary to compare the performance of each component of the final algorithm to the quality of the generated solutions. To that end, we compare the constructive method isolated, then coupled with the local search method and, finally, the complete *BVNS* algorithm. Table 2 shows the results obtained in this experiment.

Table 2. Analysis of the contribution of each component of the proposed algorithm.

Algorithm	Avg	Time (s)	Dev. (%)	#Best
GreedyConstructive	1837.80	**3.14**	0.41	1
GreedyConstructive+LS	1834.10	17.90	0.27	1
BVNS	**1829.40**	42.45	**0.00**	10

Although the constructive method isolated and the coupled with LS allows the algorithm to find 1 best solution in the instance *frb35-17-1*, *BVNS* is the one able to reach all the best solutions. It is worth mentioning that, although the constructive procedure and the local search are not able to reach more than one best solution, the small deviation indicates that they provide a promising starting point for the BVNS without being computationally demanding methods.

Finally, the BVNS is compared with the best method found in the literature for MPP, a hybrid search evolutionary algorithm, named (LS+PI) EA [19], which proposes an effective hybrid search heuristic but leveraging the combination of a greedy local search method with evolution-based heuristics. In this experiment the complete set of 35 instances is considered, and the results are depicted in Table 3.

Table 3. Comparison of the hybrid search evolutionary algorithm (LS+PI) EA and the proposed BVNS for each instance type.

Algorithm	Avg	Time (s)	Dev. (%)	#Best
BVNS	**4724.09**	444.77	**0.02**	**33**
(LS + PI) EA	5472.54	**155.29**	8.92	3

In this case, BVNS is able to reach 33 out of 35 best solutions, which highlights the efficiency of BVNS, while (LS+PI) EA is able to reach just 3 best solutions. Regarding the computing time, it can be seen that (LS+PI) EA requires less computing time, but BVNS is still satisfying the MPP constraint of requiring short computing times. The average deviation shown by BVNS indicates that, in the 2 instances in which the best solution is not found, it still remains really close to it. On the contrary, (LS+PI) EA presents a deviation of almost 9%, indicating that it is not close to the best solution. Finally, it is important to remark that the largest differences in deviation are obtained in the most complex instances, highlighting the scalability of our proposal. Analyzing these results, we can conclude that the proposed BVNS is a competitive algorithm for solving the MPP.

5   Conclusions

This research presents a BVNS approach for solving the Monitor Placement Problem (MPP) efficiently. Due to the practical applications of this problem, it is necessary to provide high quality solutions in short computing times. To that end, a BVNS algorithm is presented, where each component of the final algorithm is carefully designed to avoid being extremely computationally demanding. The specific design of the constructive procedure, the local search method and the shake procedure, allows the algorithm to be completely scalable as it can be seen in the results obtained.

A greedy constructive procedure is proposed to produce a promising initial solution. The *GreedyConstructive* starts from scratch and constructs a solution by locating monitors in those nodes which minimize the penalty in each step. Then, a local search method, coupled with a *purge* procedure to remove redundant monitors is proposed. The experimental results show how every component of the proposed algorithm has a positive effect in the final results, emerging BVNS as a competitive method for solving the MPP. Even more, the constructive procedure coupled with the local search method provides a high quality starting point which can be considered in those cases that require real-time performance.

Future lines of research comprehend the optimization of the local search and the shake movement, and discarding nodes that will not improve the solution, trying to design a faster algorithm preserving the quality achieved in this work.

References

1. Andersson, G., et al.: Causes of the 2003 major grid blackouts in north america and europe, and recommended means to improve system dynamic performance. IEEE Trans. Power Syst. **20**(4), 1922–1928 (2005)
2. Back, T., Khuri, S.: An evolutionary heuristic for the maximum independent set problem. In: Proceedings of the First IEEE Conference on Evolutionary Computation. IEEE World Congress on Computational Intelligence, pp. 531–535. IEEE (1994)
3. Borgatti, S.P., Everett, M.G.: A graph-theoretic perspective on centrality. Soc.l Netw. **28**(4), 466–484 (2006)
4. Chandu, D.P.: A parallel genetic algorithm for three dimensional bin packing with heterogeneous bins. arXiv preprint arXiv:1411.4565 (2014)
5. Crucitti, P., Latora, V., Marchiori, M.: Model for cascading failures in complex networks. Phys. Rev. E **69**(4), 045104 (2004)
6. Dinur, I., Safra, S.: On the hardness of approximating minimum vertex cover. Ann. Math. **162**, 439–485 (2005)
7. Duarte, A., Martí, R., Glover, F., Gortazar, F.: Hybrid scatter tabu search for unconstrained global optimization. Ann. Oper. Res. **183**(1), 95–123 (2011)
8. Evans, I.K.: Evolutionary algorithms for vertex cover. In: International Conference on Evolutionary Programming, pp. 377–386. Springer, Cham (1998). https://doi.org/10.1007/BFb0040790
9. Hansen, P., Mladenović, N., Todosijević, R., Hanafi, S.: Variable neighborhood search: basics and variants. EURO J. Comput. Optim. **5**(3), 423–454 (2017)
10. Hatano, N., Suzuki, M.: Quantum Annealing and Other Optimization Methods (2005).https://doi.org/10.1007/11526216
11. Holland, J.H., et al.: Adaptation in Natural and Artificial Systems: An Introductory Analysis with Applications to Biology, Control, and Artificial Intelligence. MIT Press (1992)
12. Karp, R.M.: Reducibility among combinatorial problems. In: Complexity of Computer Computations, pp. 85–103. Springer, Cham (1972). https://doi.org/10.1007/978-1-4684-2001-2_9
13. Lagraa, S., François, J.: Knowledge discovery of port scans from DarkNet. In: 2017 IFIP/IEEE Symposium on Integrated Network and Service Management (IM), pp. 935–940. IEEE (2017)

14. Lawler, E.L., Wood, D.E.: Branch-and-bound methods: a survey. Oper. Res. **14**(4), 699–719 (1966)
15. Luling, R., Monien, B.: Load balancing for distributed branch & bound algorithms. In: Proceedings Sixth International Parallel Processing Symposium, pp. 543–548. IEEE (1992)
16. Milanovic, M.: Solving the generalized vertex cover problem by genetic algorithm. Comput. Inform. **29**(6), 1251–1265 (2010)
17. Mueller-Bady, R., Gad, R., Kappes, M., Medina-Bulo, I.: Using genetic algorithms for deadline-constrained monitor selection in dynamic computer networks. In: Proceedings of the Companion Publication of the 2015 Annual Conference on Genetic and Evolutionary Computation, pp. 867–874 (2015)
18. Mueller-Bady, R., Kappes, M., Medina-Bulo, I., Palomo-Lozano, F.: Optimization of monitoring in dynamic communication networks using a hybrid evolutionary algorithm. In: Proceedings of the Genetic and Evolutionary Computation Conference, pp. 1200–1207 (2017)
19. Mueller-Bady, R., Kappes, M., Medina-Bulo, I., Palomo-Lozano, F.: An evolutionary hybrid search heuristic for monitor placement in communication networks. J. Heurist. **25**(6), 861–899 (2019)
20. Newman, M.E.: A measure of betweenness centrality based on random walks. Soc. Netw. **27**(1), 39–54 (2005)
21. Pelikan, M., Goldberg, D.E.: Hierarchical bayesian optimization algorithm. In: Scalable Optimization via Probabilistic Modeling, pp. 63–90. Springer, Cham (2006). https://doi.org/10.1007/978-3-540-34954-9_4
22. Pérez-Peló, S., Sánchez-Oro, J., Gonzalez-Pardo, A., Duarte, A.: A fast variable neighborhood search approach for multi-objective community detection. Appl. Soft Comput. **112**, 107838 (2021)
23. Sánchez-Oro, J., Pantrigo, J., Duarte, A.: Combining intensification and diversification strategies in VNS. An application to the vertex separation problem. Comput. Oper. Res. **52**, 209–219 (2014)
24. Zhang, D., Cetinkaya, E.K., Sterbenz, J.P.: Robustness of mobile ad hoc networks under centrality-based attacks. In: 2013 5th International Congress on Ultra Modern Telecommunications and Control Systems and Workshops (ICUMT), pp. 229–235. IEEE (2013)

A VNS-Based Heuristic for the Minimum Number of Resources Under a Perfect Schedule

Rachid Benmansour[1,2]([✉]) [iD]

[1] Institut National de Statistique et d'Economie Appliquée (INSEA), Rabat, Morocco
r.benmansour@insea.ac.ma
[2] LAMIH UMR CNRS 8201, Université Polytechnique Hauts de France, Valenciennes, France

Abstract. This paper presents a variable neighborhood search based heuristic to minimise the number of resources for the single-processor scheduling problem with time restrictions which is known to be NP-complete problem. In particular, we study the performance of the Basic Variable Neighborhood Search algorithm (BVNS) under the use of different initial solutions. The obtained results were compared with an exact method published in the literature. Computational results show that the proposed algorithm is efficient and effective as it can obtain optimal solutions in 95.55% of the cases in a reasonable amount of time.

Keywords: Scheduling · Mixed integer programming · Variable neighborhood search · Single server

1 Introduction

The present work studies the minimisation of the number of resources for the single-processor scheduling problem with time restrictions (*problem* ($\mathcal{B}$)). This problem differs from the single-processor scheduling problem with time restrictions (*problem* ($\mathcal{P}$)) that was studied first in [5]. In fact, in the problem ($\mathcal{P}$), there are several jobs to be scheduled on one processor shared among multiple identical resources. The processor can execute one job at a time such that during any time interval with length $\alpha > 0$ the number of jobs being executed is less than or equal to the number of external resources B. The objective is to find an optimal sequence of the jobs that minimises the makespan or C_{max}, which is the most studied objective function for ($\mathcal{P}$). This problem was shown to be NP-hard, independently, by Zhang et al. [19] and Benmansour et al. [4]. We underline here that it has been shown that ($\mathcal{P}$) is a particular case of another scheduling problem which is called the parallel machine scheduling problem with a single server (see [4,6]). An application of the problem can be found in logistics: In this situation, the processor represents a loading server in a supplier's central depot with several identical trucks. The processing time p_i of each job will be equivalent to the loading time of the client's order i. The parameter α will represent the time it takes for a truck to reach this customer, unload the goods and return back to the central depot to make another delivery. Other applications of this problem can be consulted here [7,12,18].

In the problem ($\mathscr{B}$) the desired goal is to minimize the number of identical resources used to obtain a perfect scheduling (*i.e.*, a schedule without idle time on the shared processor). Note that in a perfect schedule the makespan is equal to the sum of the processing times of the jobs. Consequently the problem to be addressed in this paper is to determine both the minimum value of B and a sequence of the jobs such that the processor works constantly until the completion of the last job. Since ($\mathscr{B}$) is NP-complete problem [2], it is often useful to solve it by metaheuristics, especially if the exact methods struggle to give satisfaction. In this work, a Variable Neighborhood Search (VNS) based algorithm [8] is proposed to solve ($\mathscr{B}$) as the proposed exact method in [2] was not able to solve optimally all the instances. Several scheduling rules are proposed to generate initial solutions for our VNS algorithm. In order to assess the performance of the algorithm, computational experiments are conducted on randomly generated instances that was previously solved by a mixed integer linear programming (MIP) formulation [2].

The remaining of this work is organized as follows. In Sect. 2, a formal definition of the studied problem is proposed. The proposed VNS-based algorithm and its various components are presented in Sect. 3. Section 4 is devoted to the presentation and the analysis of the results. Finally, Sect. 5 aims to conclude this work and discuss future research directions.

2   Description of the Problem

The problem ($\mathscr{B}$) can be formally described as follows. We are given a set of n jobs, $n \in N = \{1, 2, \ldots, n\}$, and each job i has a processing time $p_i > 0$. All the jobs must be processed on a single processor shared between several identical resources (their number B is to be determined). During its execution on the processor, the job i requires the use of one of the available resources throughout the duration of its processing. Once the resource is released by the job i, it becomes unavailable for $\alpha > 0$ units of time (*e.g.*, for maintenance purposes).

The objective is to determine the perfect schedule that minimizes the number of external resources used among the available resources. This problem can be of practical interest in the following situations:

- The main processor should not be stopped: Scheduling with no-idle time constraint can arise in real situation when the processor idle time cost is extremely high or when the processor cannot be easily started and stopped due to technological constraints [10,17].
- Setup costs of using external resources may be very high [1]. Therefore, using more external resources than necessary - to achieve the same result - is an obvious underperformance.

A Mixed Integer Linear Programming (MIP) formulation was proposed in [2] to solve optimally this problem. The MIP model was able to solve part of the problem instances optimally. In this paper we consider that B is a decision variable to be determined. It is obvious that B should be greater or equal to 2 in order to have a feasible schedule without idle times on the main processor. We assume also that there are as many resources as available jobs since, in the worst case, it would take n resources to execute all the jobs without idle time on the processor.

The problem considered in this paper was motivated by the following observation: In the problem ($\mathscr{P}$), the makespan (C_{max}) is a non-increasing function of the number of resources B used in the schedule. In other words, the optimal solution value of the problem ($\mathscr{P}$) decreases when the number of resources B increases. The following example is given for better illustration.

Illustrative Example: Let us consider an instance of the problem ($\mathscr{P}$) with 10 jobs. Each job i has a deterministic processing time p_i and let $\alpha = 100$ (see Table 1). We denote by SP the sum of the processing times of all the jobs. The example instance has been solved for different values of B. The MIP model given in [3] was used to solve these problems. The resulting objective function values are connected and plotted in Fig. 1.

Table 1. Example instance for $n = 10$.

Job i	1	2	3	4	5	6	7	8	9	10
p_i	7	14	19	71	25	27	49	31	38	38

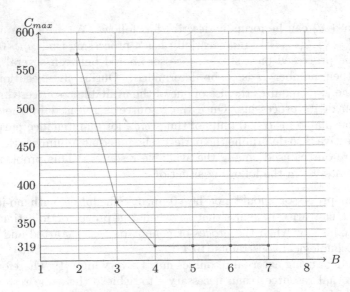

Fig. 1. The variation of C_{max} as a function of B.

From Fig. 1, we observe that for a fixed number of jobs, the value of optimal makespan decreases with the value of external resources B until it reaches the value $SP = 319$. This is an expected result since SP is a lower bound on C_{max} and the time restrictions constraint becomes no restrictive as B increases: In the extreme case where there are as many resources as there are jobs (i.e., $B = n$), a perfect schedule can be obtained by using a different resource for each job (for example the resource i can be used to process job i, $\forall i \in N$).

Figure 2 shows the optimal solution of the same instance with $B = 4$ which is the minimum number of external resources needed to get a perfect schedule.

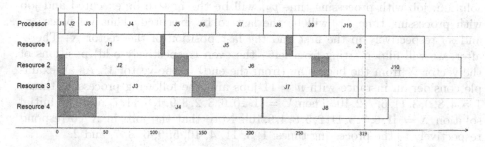

Fig. 2. A perfect schedule for the considered instance with $B = 4$.

3 The Proposed VNS-Based Algorithm

In 1997, Mladenovic and Hansen proposed a new metaheuristic method based on systematic changes in neighborhood structure. This method was called Variable Neighborhood Search (VNS) [15]. The main goal of this method is to solve complex optimization problems. Due to the few parameters it requires and its simplicity of implementation, several variants of VNS have been proposed in the literature [9]. VNS algorithm has been applied successfully to solve several problems in the literature [11,13,14,16]. Basically VNS consists of three major components: i) local search, ii) shaking procedure and iii) neighborhood change. VNS also needs an initial solution to start the search.

In this work a Basic Variable Neighborhood Search (BVNS) algorithm is proposed to solve problem ($\mathscr{B}$). The algorithm performs local searches to reach local optimum in addition to a shaking procedure to avoid getting trap in local optimum. In this work a local search procedure with best improvement strategy is used [9]. In following sections, the main components of the proposed BVNS algorithm are described.

3.1 Initial Solution

In the proposed BVNS algorithm, a solution for ($\mathscr{B}$) is represented by permutation $\pi = (\pi_1, \pi_2, \ldots, \pi_n)$, where $\pi_i \in N$ is the job executed in position $i \in N$, and p_{π_i} is its corresponding processing time on the processor.

To generate an initial solution we first tested some simple scheduling rules such as Shortest Processing Time (SPT), LPT (Longest Processing Time) and First Come First Served rule (FCFS) (*i.e.* in the order of appearance of the jobs in the instance). In a second time we proposed and tested another rule which we called *Wave Sort* or *WS* for short. The *WS* rule is by far the most efficient one. Hence to generate an initial solution X of a good quality for the problem $(\mathscr{B})$ we used the *WS* rule: First we sort the jobs in an increasing order of their processing times. Then we store the result in a vector V. Without loss of generality, we assume that we have a vector $V = [\pi_1, \pi_2, \dots, \pi_n]$ such that $p_{\pi_1} \leq p_{\pi_2} \leq \dots\dots \leq p_{\pi_n}$. According to [2] (cf. Theorem 2), in an optimal solution, job with processing time p_{π_1} will be the first to be executed and job with processing time p_{π_2} will be the last to be executed. Thus we place, π_1 and π_2 respectively in the first and the last position of the vector X. Thereafter, we will fill - starting from left - the even (respectively odd) positions of the vector X from the beginning (from the end) of the vector V. As an example consider an instance with $n = 11$ jobs and the following processing times: $1, 5, 4, 8, 9, 3, 11, 6, 7, 2, 10$. Then $V = [1, 10, 6, 3, 2, 8, 9, 4, 5, 11, 7]$ and the initial solution $X = [1, 6, 7, 3, 11, 2, 5, 8, 4, 9, 10]$. Note that the jobs in X correspond respectively to the processing times: 1, 3, 11, 4, 10, 5, 9, 6, 8, 7, and 2.

3.2 Evaluation Function

In order to evaluate a solution π, the Algorithm 1 is proposed. Given a solution $\pi = (\pi_1, \pi_2, \dots, \pi_n)$, the algorithm below returns the value of the minimal external resources B needed in order to satisfy the no-idle time constraints. We define L as the final list of the external resources used in each solution.

Algorithm 1: Algorithm to compute optimal number B given a sequence of jobs π

Data: A sequence of jobs $\pi = (\pi_1, \pi_2, \dots, \pi_n)$.
Result: Minimal number of resources B to obtain a perfect schedule given π.
Initialization: $B = 2$, $L = \{1, 2\}$;
Schedule job π_1 on the processor using resource 1;
Schedule job π_2 on the processor using resource 2;
for *i=3* **to** n **do**
    if *a resource* $r \in L$ *is available at time* $t = \sum_{s=1}^{i-1} p_{\pi_s}$ **then**
       | Use resource r to process the job i on the processor;
    end
    else
       | $B \leftarrow B + 1$;
       | $L \leftarrow L \cup \{B\}$ // add a new resource to the list L
    end
end
return B;

Initially we set $B = 2$ and $L = \{1, 2\}$ as there is always idle times on the processor with one external resource. We schedule the jobs π_1 and π_2 on the processor using respectively resources 1 and resource 2. For the following jobs $\pi_i, i \geq 3$, if there exists a resource available at time $t = \sum_{s=1}^{i-1} p_{\pi_s}$, then we should use this resource to process the job π_i on the processor. Otherwise we should add a new resource to the list L to avoid idle time on the processor. Finally the algorithm returns B.

3.3   Neigborhood Structure

A neighborhood structure is designed to move from one solution to another through the execution of a set of elementary operations. In scheduling problems we are mainly interested in changing the order of the execution of the jobs from one solution to another. We define herein the two neighborhood structures used in our algorithm. These neighborhood structures are defined by their correspond-ing operators. For the sake of simplicity, a neighborhood structure will also be referred to by the name of its corresponding generating operator.

- The insertion operator ($\mathcal{N}_1$): For a given solution, the insertion neighborhood can be obtained by removing a job from its current position and inserting it into another position at random.
- The swap operator ($\mathcal{N}_2$): The neighborhood set consists of all solutions obtained by swapping two jobs at random in the current solution.

3.4   Shaking and Local Search

In Algorithm 2, the initial solution is generated according to WS rule. In order to escape from local optima, the shaking procedure **Shake** is used to randomly gen-erate a neighboring solution π' from the kth neighbourhood of π. We apply, in the intensification phase, the best improvement local search procedure (**LSearch**) to the perturbed solution and save the best found solution to π''. The function **NChange** compares the new value of the solution π'' with the value of the solution π. If solution π'' is better than π, then **NChange** keeps this solution instead of π (*i.e.* $\pi \leftarrow \pi''$) and k is returned to its initial value 1; otherwise, it further perturbs the current incumbent solution π using the $k + 1^{th}$ neighbour-hood (*i.e.* $k \leftarrow k + 1$). In this paper the **Shake** procedure is based on insertion operator ($\mathcal{N}_1$) whereas the local search procedure **LSearch** is based on swap operator ($\mathcal{N}_2$). In order to run the BVNS algorithm, we need to assign values to two parameters k_{max} and t_{max}. This last parameter was chosen as follows: $t_{max} = 30 \times n$. Thus, more time was given to larger instances, which are the most difficult to solve by CPLEX. The value of the second parameter k_{max} was empir-ically selected: four values (5, 10, 30 and 50) were tested in solving 9 instances of various sizes. As a result, the value of k_{max} was set to 5. The pseudo code of the proposed BVNS is presented in Algorithm 2.

Algorithm 2: BVNS Algorithm

Data: π, k_{max}, t_{max}
Result: Solution π
Generate an initial solution π using WS rule;
repeat
  $\quad k \leftarrow 1$;
  $\quad$ **repeat**
    $\qquad \pi' \leftarrow$ **Shake**(π, k);
    $\qquad \pi'' \leftarrow$ **LSearch**(π');
    $\qquad$ **NChange**(π, π'', k);
  $\quad$ **until** $k = k_{max}$;
  $\quad t \leftarrow Current\_CPU\_Time$;
until $t > t_{max}$;
return π;

---

4   Computational Experiments

This section reports the computational experiments performed to compare the performance of the proposed BVNS algorithm with the MIP model proposed in [2]. The BVNS algorithm was implemented using C language on a laptop with configuration of Intel(R) Core(TM) i7-8565U CPU @ 1.80 GHz 1.99 GHz with 16.00 GB RAM. The instances on which the tests were carried out are the same as those used in [2], and can be found here. The number of jobs n was chosen from the set $\{10, 50, 100\}$. The processing times p_i ($i \in N$) were generated from integer uniform distributions in $[1, \alpha]$, where $\alpha \in \{10, 100, 1000\}$. For each combination of n and α 10 instances were randomly generated.

In the following tables we present the results obtained by the MIP model and the BVNS algorithm. In each table (Tables 2, 3, 4, 5, 6, 7, 8, 9 and 10), I represents the instance identifier ($1 \leq I \leq 10$), lb is the lower bound on the objective function value as reported in [2], and f_{MIP} is the value returned by CPLEX solver after solving the MIP model, CPU_{MIP} is the execution time of the MIP. The remaining columns, F_{best}, F_{avg} and CPU_{avg} represent respectively the best solution value, the average solution value and the CPU time of 30 runs of BVNS algorithm for instance I.

A diamond $\Diamond$ in the forth column (*i.e.* f_{MIP} values) indicates that the value obtained is not necessarily optimal because it represents the value returned by CPLEX at the end of the time limit (which is 3000 s). For BVNS the time limit was set to $t_{max} = 30 \times n$, where n is the number of the jobs in the considered instance.

***N.B.**: Contrary to what is published in [2], the value of the optimal solution for the instance with $n = 50$, $\alpha = 100$ and $I = 5$ is $f_{MIP} = 3$ (and not 4). This value is identified by the symbol ** in Table 6.*

Note that, in order to take advantage of the calculated lower bounds lb another stopping condition was added to the BVNS algorithm. Indeed, the algorithm is stopped once the value of the incumbent solution is equal to the lower bound lb of the considered instance or the time limit t_{max} is reached.

From the results obtained we can make the following comments:

- The BVNS algorithm finds the same optimal solutions as CPLEX, in less amount of time in the majority of cases.
- The lower bound is tight because in 84% of cases the value of the lower bound lb is equal to the value of the optimal solution (76 cases out of 90). In the remaining cases, the BVNS algorithm exhausts the time allocated searching for better solutions.
- Following the use of BVNS seven new best-known solutions (in bold in the f_{BVNS}-column) and 83 ties are claimed in testing the set with 90 instances. The seven new best-known solutions are optimal solutions.
- The quality of lb and WS rule values contribute to the performance of BVNS.
- The optimal value for each instance is either 3 or 4 resources. This can be explained easily as follows. In any feasible solution of the problem ($\mathscr{B}$), the sum of the processing times of each consecutive $B - 1$ jobs must be greater of equal to α (cf. [2]). Given that $p_i \in [1, \alpha]$ for each $i \in \{1, 2, \ldots, n\}$, then, on average, the sum of the processing times of the $B - 1$ consecutive jobs is equal to $S = (B - 1) \times \frac{(\alpha-1)}{2}$. Hence in order to have $S \geq \alpha$, the value of B should be greater or equal to $1 + 2\frac{2\alpha}{(\alpha-1)}$. Numerically, and given the chosen intervals $[1, 10]$, $[1, 100]$ and $[1, 1000]$, the value of B must be minimal and respectively equal to at least 3.222, 3.020 and 3.002.

Finally we can conclude that the proposed BVNS algorithm is efficient to solve the proposed instances. This is not the case of the MIP model which struggles to solve several instances, in an optimal way, even after 3000 s. The efficiency of BVNS algorithm also due to the quality of the initial solution and to the kind of instances considered (generated according to the uniform distribution).

Table 2. Results obtained for instances with $n = 10$ and $\alpha = 10$.

I	lb	MIP		BVNS		
		F_{MIP}	CPU_{MIP}	F_{best}	F_{avg}	CPU_{avg}
1	3	3	0.01	3	3	0
2	3	3	0.03	3	3	0.0001
3	3	3	0.02	3	3	0
4	3	3	0.02	3	3	0
5	3	3	0.02	3	3	0
6	3	3	0.02	3	3	0
7	3	3	0.02	3	3	0
8	4	4	0.08	4	4	0
9	3	3	0.01	3	3	0
10	3	4	0.02	4	4	300

Table 3. Results obtained for instances with $n = 10$ and $\alpha = 100$.

I	lb	MIP		$BVNS$		
		F_{MIP}	CPU_{MIP}	F_{best}	F_{avg}	CPU_{avg}
1	3	3	0.03	3	3	0
2	3	4	0.02	4	4	10
3	3	3	0.02	3	3	0
4	3	4	0.03	4	4	10
5	3	3	0.01	3	3	0
6	3	4	0.11	4	4	10
7	3	3	0.02	3	3	0
8	3	3	0.02	3	3	0
9	4	4	0.03	4	4	0
10	4	4	0.06	4	4	0

Table 4. Results obtained for instances with $n = 10$ and $\alpha = 1000$.

I	lb	MIP		$BVNS$		
		F_{MIP}	CPU_{MIP}	F_{best}	F_{avg}	CPU_{avg}
1	3	3	0.01	3	3	0
2	3	3	0.01	3	3	0
3	3	3	0.01	3	3	0
4	4	4	0.02	4	4	0
5	3	3	0.02	3	3	0
6	3	3	0.01	3	3	0
7	3	3	0.02	3	3	0
8	3	3	0.01	3	3	0
9	3	3	0.01	3	3	0
10	4	4	0.02	4	4	0

Table 5. Results obtained for instances with $n = 50$ and $\alpha = 10$.

I	lb	MIP		$BVNS$		
		F_{MIP}	CPU_{MIP}	F_{best}	F_{avg}	CPU_{avg}
1	3	3	1.58	3	3	0.0014
2	3	3	19.41	3	3	0.001
3	3	3	2.2	3	3	0.001
4	3	3	7.61	3	3	0.0012
5	3	3	2.61	3	3	0.001
6	3	3	3.59	3	3	0.0009
7	3	3	3.48	3	3	0.0011
8	3	3	3.69	3	3	0.0009
9	3	3	3.3	3	3	0.0009
10	3	4	0.75	4	4	1500

Table 6. Results obtained for instances with $n = 50$ and $\alpha = 100$.

I	lb	MIP		BVNS		
		F_{MIP}	CPU_{MIP}	F_{best}	F_{avg}	CPU_{avg}
1	3	3	8.27	3	3.1	0.001
2	3	3	235.03	3	3	0.001
3	4	4	0.03	4	4	0.001
4	3	3	1.77	3	3	0.0005
5	3	3**	443.28	3	3	0.001
6	4	4	0.08	4	4	0.001
7	3	4	36.66	4	4	1500
8	3	4	11.02	4	4	1500
9	4	4	0.06	4	4	0.001
10	3	4	38.39	4	4	1500

Table 7. Results obtained for instances with $n = 50$ and $\alpha = 1000$.

I	lb	MIP		BVNS		
		F_{MIP}	CPU_{MIP}	F_{best}	F_{avg}	CPU_{avg}
1	3	3	1182.61	3	3	0.0007
2	3	4$^\diamond$	3000	4	4	1500
3	3	3	734.09	3	3	0.0006
4	4	4	0.06	4	4	0.001
5	4	4	0.03	4	4	0.0007
6	3	3	5.77	3	3	0.0003
7	3	4	6.86	4	4	1500
8	4	4	0.06	4	4	0
9	4	4	0.08	4	4	0.0007
10	3	4$^\diamond$	3000	**3**	3	0.0007

Table 8. Results obtained for instances with $n = 100$ and $\alpha = 10$.

I	lb	MIP		BVNS		
		F_{MIP}	CPU_{MIP}	F_{best}	F_{avg}	CPU_{avg}
1	3	3	12.89	3	3	0.0055
2	3	4	3.78	4	4	3000
3	3	3	17.28	3	3	0.006
4	3	3	22.14	3	3	0.006
5	3	3	741.95	3	3	0.006
6	3	3	26.99	3	3	0.006
7	3	3	20.8	3	3	0.006
8	3	3	104.28	3	3	0.0065
9	3	3	17.78	3	3	0.006
10	3	3	38.58	3	3	0.006

Table 9. Results obtained for instances with $n = 100$ and $\alpha = 100$.

I	lb	MIP		$BVNS$		
		F_{MIP}	CPU_{MIP}	F_{best}	F_{avg}	CPU_{avg}
1	3	$4^\diamond$	3000	**3**	3	0.0076
2	4	4	0.23	4	4	0.0068
3	4	4	0.13	4	4	0.0062
4	4	4	0.2	4	4	0.0056
5	3	$4^\diamond$	3000	**3**	3	0.0054
6	3	4	568.3	4	4	3000
7	3	$4^\diamond$	3000	**3**	3	0.007
8	3	$4^\diamond$	3000	**3**	3	0.007
9	3	$4^\diamond$	3000	4	4	3000
10	3	$4^\diamond$	3000	4	4	3000

Table 10. Results obtained for instances with $n = 100$ and $\alpha = 1000$.

I	lb	MIP		$BVNS$		
		F_{MIP}	CPU_{MIP}	F_{best}	F_{avg}	CPU_{avg}
1	3	$4^\diamond$	3000	4	4	73.0017
2	4	4	0.22	4	4	0.0061
3	4	4	0.2	4	4	0.0062
4	3	3	50.45	3	3	0.0058
5	3	$4^\diamond$	3000	**3**	3	0.0059
6	3	$4^\diamond$	3000	3	3	0.0062
7	3	3	24	3	3	0.0056
8	3	$4^\diamond$	3000	4	4	73.0024
9	3	4	4.7	4	4	73.0014
10	4	4	0.22	4	4	0.0069

5   Conclusion

We introduce a new problem related to the single processor scheduling problem with time restrictions. In this problem the objective is to find the minimum number of external resources in order to have a perfect schedule on the server. We propose a VNS-based heuristic to solve large instances of the problem. Also we compared the results obtained with those previously published in the literature and obtained thanks to the resolution of a MIP model. In the light of the results obtained, it appears that BVNS algorithm is by far more efficient than the exact MIP model in terms of time and quality of the solution. However, we believe that this experimental study is insufficient and partial. For this reason, we believe that it is necessary to study other instances, of larger size and possibly generated according to other probability distributions, in order to draw more general conclusions on the efficiency of the proposed algorithm.

References

1. Allahverdi, A.: A survey of scheduling problems with no-wait in process. Eur. J. Oper. Res. **255**(3), 665–686 (2016)
2. Benmansour, R., Braun, O.: On the minimum number of resources for a perfect schedule. Cent. Eur. J. Oper. Res. **31**, 1–14 (2022)
3. Benmansour, R., Braun, O., Artiba, A.: Mixed integer programming formulations for the single processor scheduling problem with time restrictions. In: CIE 45: 2015 International Conference on Computers and Industrial Engineering (2015)
4. Benmansour, R., Braun, O., Hanafi, S.: The single-processor scheduling problem with time restrictions: complexity and related problems. J. Sched. **22**(4), 465–471 (2019)
5. Braun, O., Chung, F., Graham, R.: Single-processor scheduling with time restrictions. J. Sched. **17**(4), 399–403 (2014)
6. Brucker, P., Dhaenens-Flipo, C., Knust, S., Kravchenko, S.A., Werner, F.: Complexity results for parallel machine problems with a single server. J. Sched. **5**(6), 429–457 (2002)
7. Elidrissi, A., Benmansour, R., Benbrahim, M., Duvivier, D.: Mathematical formulations for the parallel machine scheduling problem with a single server. Int. J. Prod. Res. **59**(20), 6166–6184 (2021)
8. Hansen, P., Mladenović, N.: Variable neighborhood search: Principles and applications. Eur. J. Oper. Res. **130**(3), 449–467 (2001)
9. Hansen, P., Mladenović, N., Pérez, J.A.M.: Variable neighbourhood search: methods and applications. Ann. Oper. Res. **175**(1), 367–407 (2010)
10. Kacem, I., Kellerer, H.: Approximation algorithms for no idle time scheduling on a single machine with release times and delivery times. Discret. Appl. Math. **164**, 154–160 (2014)
11. Karakostas, P., Sifaleras, A., Georgiadis, M.C.: Basic VNS algorithms for solving the pollution location inventory routing problem. In: Sifaleras, A., Salhi, S., Brimberg, J. (eds.) ICVNS 2018. LNCS, vol. 11328, pp. 64–76. Springer, Cham (2019). https://doi.org/10.1007/978-3-030-15843-9_6
12. Kim, H.J., Lee, J.H.: Scheduling uniform parallel dedicated machines with job splitting, sequence-dependent setup times, and multiple servers. Comput. Operat. Res. **126**, 105115 (2021)
13. Krim, H., Benmansour, R., Duvivier, D., Artiba, A.: A variable neighborhood search algorithm for solving the single machine scheduling problem with periodic maintenance. RAIRO-Oper. Res. **53**(1), 289–302 (2019)
14. Mladenović, N., Alkandari, A., Pei, J., Todosijević, R., Pardalos, P.M.: Less is more approach: basic variable neighborhood search for the obnoxious p-median problem. Int. Trans. Oper. Res. **27**(1), 480–493 (2020)
15. Mladenović, N., Hansen, P.: Variable neighborhood search. Comput. Oper. Res. **24**(11), 1097–1100 (1997)
16. Pei, J., Mladenović, N., Urošević, D., Brimberg, J., Liu, X.: Solving the traveling repairman problem with profits: A novel variable neighborhood search approach. Inf. Sci. **507**, 108–123 (2020)

17. Tanaka, S.: An exact algorithm for single-machine scheduling without idle time. In: Third Multidisciplinary International Scheduling Conference: Theory and Applications (MISTA2007). pp. 314–317 (2007)
18. Torjai, L., Kruzslicz, F.: Mixed integer programming formulations for the biomass truck scheduling problem. CEJOR **24**(3), 731–745 (2016)
19. Zhang, A., Chen, Y., Chen, L., Chen, G.: On the np-hardness of scheduling with time restrictions. Discret. Optim. **28**, 54–62 (2018)

BVNS for Overlapping Community Detection

Sergio Pérez-Peló[(⊠)] [iD], Jesús Sánchez-Oro [iD], Antonio González-Pardo [iD],
and Abraham Duarte [iD]

Dept. Computer Science, Universidad Rey Juan Carlos,
28933 Móstoles, Madrid, Spain
{sergio.perez.pelo,jesus.sanchezoro,antonio.gpardo,
abraham.duarte}@urjc.es

Abstract. Nowadays, social networks are one of the most important
sources of information available on the Internet, since new users and
relationships between them emerge every day in this type of networks.
For this reason, it is important to have procedures and mechanisms to
obtain, process and analyze the information extracted from them and
transform it into useful data. This situation has given rise to new hard
combinatorial optimization problems related to social networks, such as
influence analysis, sentimental analysis or polarization. All these topics
are grouped under the research field of Social Networks Analysis (SNA).
In this paper, we focus on one of these topics: the Community Detection
Problem (CDP). Specifically, we will deal with a variant of the CDP
known as the Overlapping Community Detection Problem (OCDP), in
which the same user can be assigned to more than one community simultaneously, which cannot occur in the classical Community Detection
Problem. The problem is approached from a heuristic point of view,
applying a Greedy Randomized Adaptive Search Procedure (GRASP)
combined with a Basic Variable Neighborhood Search (BVNS) algorithm.
The proposal is compared with the best method found in the literature, a
Density Peaks based algorithm. Synthetic instances are used to evaluate
the performance of the proposal. To analyze the quality of the obtained
solutions, an evaluation metric that has been adapted from the well-
known modularity metric has been used: the overlapping modularity.

Keywords: GRASP · VNS · Overlapping Community Detection ·
Heuristics · Optimization

1 Introduction

In recent years, the information that can be found in Internet has exponentially
grown. Every day individual users and companies obtain part of this information to use it in their own profit. Social networks (SN) are one of the biggest

This research was funded by "Ministerio de Ciencia, Innovación y Universidades" under
grant ref. PGC2018-095322-B-C22, "Comunidad de Madrid" and "Fondos Estructurales" of European Union with grant refs. S2018/TCS-4566, Y2018/EMT-5062.

A. Sleptchenko et al. (Eds.): ICVNS 2022, LNCS 13863, pp. 27–41, 2023.
https://doi.org/10.1007/978-3-031-34500-5_3

sources of information that can be used today in the Internet. Every day, new users and relations among them born in the context of social networks. These users are typically very different among them in terms of gender, age or provenance, providing a high volume of information from different social spectra. This characteristic has attracted the interest of researchers from different knowledge areas, such as psychology, marketing or data science.

The problems derived from the study of social networks can be summarized into the Social Network Analysis (SNA) research field. This field groups different real life problems that are related to the social network context. One of the most interesting problems is the analysis of the relations among different users of certain social network with the aim of grouping them into communities. In this work, the Community Detection Problem (CDP) is tackled. More specifically, the studied variant is the Overlapping CDP (OCDP), in which a user can be assigned to more than one community simultaneously. This problem models a more realistic behavior than the classical CDP, given that in modern social networks a user can be grouped in different communities depending on different aspects, such as its interests, the people that it is related to, the places that it visits, etc.

To solve this problem, our proposal consists of an approach that applies Greedy Randomized Search Adaptive Search Procedure (GRASP) [6] as a constructive method to provide initial solutions to the Basic Variable Neighborhood Search (BVNS) [10] framework. These are two well-known metaheuristics that have been proven to be useful for solving different $\mathcal{NP}$-hard problems.

The rest of the work is organized as follows: Sect. 2 contains the formal description of the problem to be addressed, as well as the metric that is applied to evaluate the quality of a given solution. In Sect. 3 the algorithmic approach applied for solving the problem is explained. Section 4 summarizes the experiments performed, devoted to evaluate the quality of the proposal. Finally, Sect. 5 enumerates the conclusions derived from the work, as well as the future work lines.

2 Problem Description

A graph $G = (V, E)$ can be used to model a Social Network. In this representation, the set of vertices V corresponds to the users of the network (with $|V| = n$), while the set of edges E is conformed with tuples $(u, v) \in E$, with $u, v \in V$, representing that there exist a relationship between users u and v. In this work, if there exists a relation between user u and user v, then it is assumed that the relation between user v and user u is also given, it is, the relations are bidirectional.

The main difference between the Overlapping Community Detection Problem (OCDP), tackled in this work, and the classical CDP is that, in the present variant, a user can be assigned to more than community at once.

A community can be defined as a subset of users (and their relationships), it is, a subgraph of the original one. More formally, a community C_k is defined as $C_k = (V_k, E_k)$, where $V_k \subset V$ is the set of users assigned to the community and E_k is the set of the relations between users in V_k. Mathematically, E_k can be defined as $E_k = \{(u, v) \in E : u, v \in V_k\}$.

The task to be solved in the context of OCDP is to assign users to one or more communities. There exists different metrics that evaluates the quality of a given community, but all of them have a common goal: to produce subgraphs whose nodes are densely connected among them and sparsely connected to nodes in other subgraphs.

A solution S for the OCDP is modeled as a set of k communities, it is, $S = \{C_1, C_2, ..., C_k\}$, where $1 \leq k \leq n$, indicating the number of found communities. In this problem, the value of k is not fixed, so it must be determined by the developed algorithm. A feasible solution is reached when all nodes in the network are associated to, at least, one community. As it has been stated, in the OCDP a user can belong to more than community at the same time. It means that certain node u can be simultaneously in community C_i and community C_j, with $1 \leq i, j \leq k$.

The overlapping feature makes the most extended metrics used in the CDP context not suitable for the OCDP. However, they can be modified to meet the particular circumstances of the OCDP. In this sense, modularity [19] is one of the most extended metrics in the context of CDP. To evaluate and compare the solutions obtained for the OCDP, an adaptation of this metric proposed in [15] is used. In this adaptation, the fact that a node can belong to more than one community at the same time is taken into account. For a single solution, the value of the modularity is calculated as the average sum of the modularity of each detected community. More formally,

$$M_O(S) = \frac{\sum\limits_{i=1}^{\mathcal{I}} M_O(C_i)}{\mathcal{I}} \tag{1}$$

where $\mathcal{I}$ is the number of detected communities in a solution. For a certain community $C_i = (V_i, E_i)$, the overlapping modularity $M_O(C_i)$ is evaluated as:

$$M_O(C_i) = \frac{1}{|V_i|} \sum_{u \in V_i} \frac{|E_{C_i \leftarrow}(u)| - |E_{C_i \rightarrow}(u)|}{d_u \cdot s_u} \cdot \frac{|E_i|}{\frac{|V_i| \cdot (|V_i| - 1)}{2}} \tag{2}$$

where $E_{C_i \leftarrow}(u)$ represent the set of edges that connect nodes belonging to the same community (intra-community edges) with an endpoint in u, $E_{C_i \rightarrow}(u)$ is the set of edges connecting nodes belonging to different communities with an endpoint in u, d_u denotes the degree of node u and s_u represents the total number of communities which node u is assigned to. Since the size of each community could be rather different, the difference between intra and inter-community edges is multiplied by the ratio between the number of edges that are actually present in the community ($|E_i|$), and the number of edges that an ideal community represented by a complete graph would have ($\frac{|V_i| \cdot (|V_i| - 1)}{2}$).

For the sake of clarity, in Fig. 1a an example network with 8 nodes and 10 edges is shown.

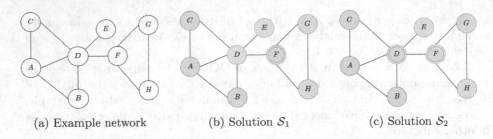

(a) Example network   (b) Solution S_1 (c) Solution S_2

Fig. 1. Example network with 8 nodes and 10 edges and two different feasible solutions for the OCDP. Figure 1a shows a network with 8 nodes and 10 edges. Figure 1b depicts a solution S_1 conformed with three communities and one overlapping node. Figure 1c exposes a different solution S_2 consisting of three communities and two overlapping nodes.

Figure 1b shows a solution S_1 with three communities with defined as $S_1 = \{C_1, C_2, C_3\}$, where $C_1 = \{A, B, C\}, C_2 = \{D, E, F\}$ and $C_3 = \{F, G, H\}$.

In this solution, only the node F is overlapped. It means that it belongs to more than one community at the same time. The modularity value for this solution is calculated as the sum of the modularity values for each community, divided by the number of found communities. Then, the modularity for each community is $M_O(C_1) = 0.19$, $M_O(C_2) = 0.34$, and $M_O(C_3) = 0.52$, resulting in a total modularity of the solution $M_O(G, S_1) = 0.35$.

Figure 1c depicts a solution S_2 defined as $S_2 = \{C_1, C_2, C_3\}$, where $C_1 = \{A, B, C, D\}, C_2 = \{D, E, F\}$ and $C_3 = \{F G, H\}$, with three different communities and two overlapping nodes: D and F.

Similarly, the modularity value of this solution is evaluated as $M_O(C_1) = 0.52$, $M_O(C_2) = 0.41$, and $M_O(C_3) = 0.52$, resulting in $M_O(G, S_2) = 0.48$.

In terms of modularity, it can be said that solution S_2 is better than solution S_1, given that the higher value of modularity, the better, because a larger modularity value indicates a better community structure in the reached solution.

The goal in the OCDP context is to find a solution $S^\star$ that maximizes the objective function value, i.e., a solution with the maximum $M_O(S^\star)$ value. In mathematical terms,

$$S^\star \leftarrow \arg\max_{S \in \mathbb{S}} M_O(G, S) \tag{3}$$

where $\mathbb{S}$ is the set of all possible solutions that can be reached by solving the OCDP for a certain network G.

In recent years, the relevance of social networks has become increasingly evident in our society. Every day more and more users connect to these kind of services To socialize, spend leisure time or get informed. In social networks can be found not only individual users, but also companies that have included these services in their business models. This results in a huge amount of data that can be analyzed to obtain relevant information.

For this reason, researchers from different knowledge areas has shown interest in social networks and the extraction of knowledge from them. The research area dedicated to studying and extracting knowledge from social networks is known as Social Network Analysis (SNA) [1]. The processes and methods included in this field has been demonstrated to be applicable to different domains, such as cybersecurity [7], marketing [26] or politics [13].

In the context of the OCDP [28], several approaches have been proposed. Most of them are adaptations of algorithms that solve the classical Community Detection Problem to deal with the operlapping possibility. Example of these adaptations can be found at [14], where a Clique-Percolation-Method (CPM) has been adapted. Another work that adapts a classical algorithm to solve the OCDP is found in [9], where the well-known Label Propagation algorithm is adapted. Other approaches apply greedy algorithms [27] or non-negative matrix factorization [30].

In the context of heuristics and metaheuristics, the best proposal found is the one proposed by Xu et al. [29], where an extended adaptive version of the Density Peak algorithm [23], named EADP, is applied to solve the OCDP. This algorithm takes into account the distance between any pair of nodes (computed as the sum of nodes between each node) to detect overlapped communities. The algorithms tries to find the center of the communities using the Density Peak algorithm.

3   Algorithmic Approach

In CDP, there exist a real necessity of develop fast algorithms that process all the amount of information extracted from social networks. This is because these kind of networks are continuously evolving, which implies that, if an algorithm takes a long time to provide results, maybe these results are outdated with respect to the current status of the network. This casuistry causes the exact algorithms not to be suitable when dealing with real-life networks, mainly due to the large extent of the solution space. In this context, heuristic algorithms emerges as a good option to deal with this kind of problems, by sacrificing the guarantee of optimality in favor of a reduced computing time. Heuristic algorithms have another main disadvantage: they can eventually get trapped in local optimum when exploring a region of the solution space. To avoid this problem, metaheuristic algorithms are developed with the aim of guiding heuristics during the traversal of the solution space in a more intelligent way. In this work, the Variable Neighborhood Search (VNS) metaheuristic [10] is applied for solving the OCDP. VNS is able to escape from local optima by performing systematic neighborhood changes until no improvement is found after a certain number of changes. This metaheuristic has been previously applied for solving different $\mathcal{NP}$-hard problems, obtaining successful results in them. Examples of its application can be found at [2, 16, 18, 21, 22].

The success of this methodology has lead to several variants that can be classified attending to the balance between intensification and diversification of the obtained solutions. If we focus on intensification, Variable Neighborhood Descent (VND) [4,24] performs deterministic neighborhood changes. If diversification is a better option for the incumbent problem, Reduced VNS (RVNS) [12] is a good choice, given that it performs stochastic neighborhood changes, providing more diverse solutions. Alternatively, if what is sought is a balance between intensification and diversification, Basic VNS (BVNS) [3] provides a good framework that combines stochastic and deterministic neighborhood changes, arising a compromise between intensification and diversification. New variants of the methodology have been proposed by researchers in recent years, such as Variable Neighborhood Decomposition Search (VNDS) [11], General VNS (GVNS) [25] or parallel approaches such as [8].

In this work, we apply the BVNS variant for solve the OCDP. The compromise between intensification and diversification in the context of OCDP allows to explore more regions of the solution space without loss of quality regarding at the objective function, as it will be experimentally tested in Sect. 4. Algorithm 1 shows the pseudo-code of the BVNS framework.

Algorithm 1. $BVNS(S, k_{\max})$

1: $k \leftarrow 1$
2: **while** $k \leq k_{\max}$ **do**
3:     $S' \leftarrow Shake(S, k)$
4:     $S'' \leftarrow Improve(S')$
5:     $k \leftarrow NeighborhoodChange(S, S'', k)$
6: **end while**
7: **return** S

The algorithm starts from an initial solution denoted with S. In the VNS framework, this initial solution can be generated in different ways. It can be generated totally at random, using a greedy approach or apply algorithms that construct solutions in a more sophisticated way. In this work, the initial solution is generated by using a whole metaheuristic: Greedy Randomized Adaptive Search Procedure (GRASP). This initial construction phase is explained in Sect. 3.1. The BVNS algorithm receives a second input parameter, k_{max}, that indicates the maximum neighborhood that is going to be explored during the search. This parameter is typically small, with the aim of avoiding the exploration of totally different solutions in each iteration of the algorithm, degenerating in a multistart approach.

Starting from the first neighborhood, $k = 1$ (step 1), BVNS iterates until the last neighborhood (defined by k_{max}) is reached (steps 2–6). In each iteration, a shake method is applied to the solution S, generating a perturbed solution S'. This shake procedure is explained in detail in Sect. 3.2. Once the perturbed solution is generated, an improvement method is applied to S' with the objective of reaching a local optimum of the perturbed solution. In this work, a local search procedure is applied to the solution. The solution S'' resulting after applying the local search procedure is passed as a parameter to the *Neighborhood Change* method (step 5). This method is in charge of selecting the next neighborhood to be explored.

More specifically, if an improvement is found, then the neighborhood change method restarts the search from the first neighborhood ($k = 1$). Otherwise, the next neighborhood is explored, and the search continues from $k = k + 1$. The algorithm stops when no improvement is found in any of the neighborhoods, it is, when k_{max} is reached, returning the best solution found during the search.

3.1 Initial Solution Generation

As it was aforementioned, the initial solution for the BVNS algorithm is generated using a whole metaheuristic: GRASP [5,6]. This metaheuristic consists of two well differentiated phases: the construction and the improvement phase. In the former, a high quality solution is generated from scratch. In the latter, the objective is to reach a local optimum with respect to certain neighborhood, starting from the constructed solution.

To generate a high-quality initial solution, it is mandatory to define a criterion that drives to select the most appropriate nodes to conform a community. In this sense, GRASP makes use of a greedy criterion that allows to assign a punctuation to each node. In this work, the selected greedy criterion is the PageRank [20] metric. This metric was originally proposed with the aim of ranking the relevance of a web page on the Internet, using their incoming links. In the context of OCDP, the larger the PageRank value, the more relevant the node is. Given that the networks that are being solved in this work are static ones, the PageRank value associated to a node only needs to be calculated once, considerably reducing the computational effort. Algorithm 2 shows the pseudo-code of the proposed constructive procedure.

The algorithm starts from an empty solution S (step 1) in which the nodes are not assigned to any community. Then, a candidate list CL is built. This list is composed by all the nodes belonging to the network G (step 2). GRASP algorithm is executed until the constraint of the OCDP is satisfied: all nodes must be assigned to, at least, one community. It means that the algorithm iterates until there are no more nodes available to be selected in the CL (steps 3–13).

The next node to form a new community is selected in each iteration. To do this, every node u in the graph is evaluated following the PageRank function $g(u)$, in order to determine the convenience of starting the community from node u. Attending to this metric, the minimum (g_{min}) and maximum (g_{max}) values are computed (steps 4–5). These two values allow to calculate a threshold μ (step 6) that is in charge of limiting the minimum value of PageRank that must have associated a node to be considered in the *Restricted Candidate List* (RCL). This RCL is devoted to provide certain randomness to the method, since the RCL will include not only the best candidate, but a group of high quality candidates. These candidates are conditioned by an α parameter that ranges from 0 to 1. The closer the value to 0, the more random the algorithm is, given that more candidates are included into the RCL. On the contrary, the closer the value to

Algorithm 2. *Construction*$(G = (V, E), \alpha)$

1: $S \leftarrow \emptyset$
2: $CL \leftarrow V$
3: **while** $CL \neq \emptyset$ **do**
4:     $g_{\min} \leftarrow \min_{v \in CL} g(v)$
5:     $g_{\max} \leftarrow \max_{v \in CL} g(v)$
6:     $\mu \leftarrow g_{\max} - \alpha \cdot (g_{\max} - g_{\min})$
7:     $RCL \leftarrow \{v \in CL : g(v) \geq \mu\}$
8:     $v \leftarrow Random(RCL)$
9:     $L \leftarrow DMF(v, G)$
10:     $CL \leftarrow CL \setminus L$
11:     $C \leftarrow InducedSubgraph(L, G)$
12:     $S \leftarrow S \cup \{C\}$
13: **end while**
14: **return** S

1, the more greedy algorithm is, given that only nodes with a PageRank value near to $g_{\max}$ are included in the RCL.

Once the RCL is built, a node u is randomly selected from it (step 8), and this node is considered the new community origin. To determine which nodes must be assigned to the community under construction the *DMF* (*Dynamic Membership Function*) algorithm is applied. Its implementation is explained in detail in [17]. Fundamentally, the algorithm is based in a breadth-first search, and iteratively adds nodes to the community under construction in such a way that included nodes improve the ratio between intra-community and inter-community edges. The traversal stops when there are no new nodes that satisfies this condition. Given that a node already included in a certain community can also be added to a different one if it improves the aforementioned ratio. This feature makes the algorithm suitable to provide an initial solution for the OCDP. Nodes that will be included in the new community are removed from the CL, with the aiming of not taking them into account as starting nodes for new communities (step 10), as this may cause the algorithm to cycle. Then, the community C conformed by the induced graph formed with the selected vertices (and the edges with an endpoint in them) is generated (step 11) and included in the solution S (step 12). Finally, the constructed solution S is returned (step 14).

After an initial solution is built, an improvement procedure is applied to reach a local optimum with respect to the explored neighborhood. The improvement procedure applied in this work is a local search procedure.

The move operator considered in the search corresponds to the addition of a vertex v to a new community C_j different from the original community that the node belongs to (C_i). At the same time, it is evaluated the profit of removing the vertex from its current community and the one obtained if it is maintained in both of them (producing an overlapping state for the current node). To decide if a node v is maintained or removed from its original community, the percentage of inter-community and intra-community edges with respect to its community are compared. Specifically, the movement is performed if and only if the percentage of inter-community edges minus the percentage of intra-community edges of node u in the current community is greater than a given threshold τ, which is a parameter of the local search (see Sect. 4 for a detailed analysis of the effect of this parameter in the procedure). Otherwise, u stays in the current community C_i but it is also incorporated in C_j. Given the definition of a good community structure, if a node has more edges to other communities than to nodes in the same one, then it should not belong to that community, since it is more related to nodes in other groups. The proposed movement is based on that idea. More formally, the move operator can be defined as:

$$Move(u, C_i, C_j, \tau) = \begin{cases} C_i \leftarrow C_i \setminus \{u\} \\ C_j \leftarrow C_j \cup \{u\} \end{cases} \text{ if } \frac{E_{\rightarrow}(u, C_i)}{d_u} - \frac{E_{\leftarrow}(u, C_i)}{d_u} > \tau \\ C_j \leftarrow C_j \cup \{u\} \quad \text{otherwise} \end{cases} \tag{4}$$

The neighborhood that will be explored by the local search procedure is conformed with all solutions that can be reached by making a single move starting from S. Mathematically,

$$N(S) = \{S' \leftarrow Move(u, C_i, C_j, \tau) \quad \forall u \in V, \quad \forall C_i, C_j \in S : u \in C_i \wedge u \notin C_j\} \tag{5}$$

Another element required to define a local search is the manner in which the neighborhood is going to be explored. Traditionally, two main strategies have been considered: best and first improvement. Following a best improvement strategy, the whole neighborhood is explored, performing the best movement found, it is, the movement that leads to the solution with the best objective function value in the neighborhood under exploration. When a first improvement strategy is followed, the movement performed is the first one that leads to a better solution in the incumbent neighborhood. In the context of OCDP, the evaluation of a solution after a move has been performed is a time-demanding task. For this reason, a first improvement strategy is selected, with the aim of reducing the computing time that the local search procedure requires.

In each iteration of the algorithm, a solution is replaced by other if and only if a neighbor improves the objective function value. It is, the acceptance criterion to move to a new solution is that it has been improved, or, in other words, when a local optimum is reached.

3.2 Shake Method

Once a local optimum is reached in the context of VNS framework. a perturbation method (known as *Shake* method) is executed to escape from the local optimum found. To do this, the method performs modifications over the found solution, reaching a different one in its neighborhood. These modifications can be defined by a movement. In the context of the OCDP, the movement consists of a removal process in which a node is unassigned from all communities that it belongs to. Then, the move is defined as:

$$Move(S, u, C_i) = V_i \leftarrow V_i \setminus \{u\} \text{ and } E_i \leftarrow E_i \setminus \{(u,v) \in E_i : v \in V_i\}, \\ for \ 1 \le i \le \mathcal{I} : u \in V_i \qquad (6)$$

This move allows to define the neighborhood of a solution, that is compound by the set of solutions that can be reached by performing it. More formally,

$$N(\mathcal{S}) = \{\mathcal{S} \leftarrow Move(S, u, C_i) : \forall v \in V \setminus C_i \wedge 1 \le i \le \mathcal{I}\} \qquad (7)$$

Given this definition, the neighborhood $N_k(\mathcal{S})$ is conformed with all solutions that can be reached when performing k consecutive movements over S. It is important to note that generated solutions are not feasible at this point, given that there exists nodes that are not assigned to any community. To solve this, a post-processing method is needed to be applied with the aim of recovering the feasibility of the perturbed solutions. To do this, a greedy approach is followed, selecting the most suitable community for a node. To decide which community is better for a node, the communities are traversed, finding the one with has a higher ratio of common edges / degree of node under evaluation. Then the node is assigned to this community.

It is important to remark that the solutions obtained after the application of *Shake* and reparation methods are not necessarily local optima. In fact, they are usually worse in terms of the objective function than the original one. Nevertheless, the main objective of *Shake* method is to escape from local optima and explore a different region of the search space. For this reason, the local search procedure defined in Sect. 3.1 is applied to locally optimize all the perturbed solutions.

3.3 Neighborhood Change

In order to select the next neighborhood that must be explored, the Neighborhood Change method is executed. This method usually takes three input parameters: the current neighborhood being explored (k), the best solution found so far and the candidate solution to be evaluated. The pseudo-code of the Neighborhood Change method is shown in Algorithm 3. Initially, if the candidate solution outperforms the best one, the latter is replaced by the former, and the search is restarted from the first neighborhood (steps 1–3). Otherwise, the next neighborhood is explored (step 5).

The whole VNS framework is executed until the Neighborhood Change method returns a k value equal to k_{max}, moment in which the algorithm is considered finished.

Algorithm 3. *NeighborhoodChange(S*, S, k)*

1: **if** $M_O(S^*) < M_O(S)$ **then**
2:     $S^* \leftarrow S$
3:     $k \leftarrow 1$
4: **else**
5:     $k \leftarrow k + 1$
6: **end if**
7: **return** k

---

4 Experiments and Results

In this section, the experiments carried out to evaluate the quality of the proposal are presented. In these experiments a set of 57 synthetic LFR networks have been solved. All the algorithms have been executed in an AMD Ryzen 5 3600 AM4 core (3.6 GHz) with 16 GB RAM. The proposed algorithm is implemented using Java 9 while the source code of the state-of-the-art method [29] is implemented in Matlab. For the experimental phase, the experiments have been divided in two different steps: preliminary and final experiments. Preliminary experiments are devoted to adjust the parameters of the proposal, while the final ones make a comparison with the best method found in the literature.

For all the experiments, the following metrics are reported: Avg., the average overlapped modularity obtained with the algorithm in the experiment; Dev.(%), the average deviation with respect to the best solution found in the experiment; Time (s), the total computing time required by each algorithm measured in seconds; and #Best, the number of times that an algorithm matches the best solution found during the experiment.

4.1 Preliminary Experiments

In this phase, the experiments are performed over a subset of instances (30 out of 57) with the aim of avoiding the overfitting of the algorithm. The first experiment is performed to evaluate the best value of the α parameter in GRASP algorithm (Sect. 3.1). Specifically, the tested values have been 0.25, 0.50, 0.75 and *RND*, where *RND* represents a random value in the range [0,1] for each iteration of the algorithm. The idea behind the selection of these values is to traverse the whole range of possible behaviors of GRASP: from a mostly greedy approach to a mostly random one. The constructive algorithm is executed for 100 independent iterations, retrieving the best solution found for each instance under evaluation. Table 1 exposes the obtained results for each α value. As it can be seen, the configuration with an alpha value $\alpha = 0.25$ is able to obtain the best results in a low computing time. It means that the more greedy approach, the better. The $\alpha = 0.5$ configuration is the second best configuration. These two configurations obtain the lower value of average deviation, which means that they are close to the best value when they are not able to reach it. Therefore, it can be derived that the configuration with $\alpha = 0.25$ is the best one for the constructive procedure.

Table 1. Comparison of the average results obtained by the GRASP algorithm with different α parameter values. Best results are highlighted with bold font.

α	Avg	Dev. (%)	Time(s)	#Best
0.25	**0.3282**	**1.58**	**33.25**	**21**
0.50	0.3258	1.84	33.79	8
0.75	0.3147	5.20	34.70	11
RND	0.3241	2.82	34.14	10

The second performed experiment is devoted to test the best value to the τ parameter in the context of the local search procedure (see Sect. 3.1). Table 2 shows the obtained results.

Table 2. Comparison of the average results obtained by the local search procedure with different values for the τ parameter. Best results are highlighted with bold font.

τ	Avg	Dev. (%)	Time (s)	#Best
0.2	**0.3328**	**0.43%**	**33.60**	**28**
0.3	0.3327	0.45%	33.70	24
0.4	0.3326	0.85%	34.90	18
0.5	0.3321	1.10%	35.04	18

As it can be seen, the best configuration corresponds to the configuration with a $\tau = 0.2$ value. It implies that, the lower difference between intra-community and inter-community edges in a community, the better. This configuration of the algorithm reaches the best solution in 28 out of 30 instances, with a low value of deviation in average (0.43%) when it is not able to find the best solution.

Finally, the last experiment is devoted to select the best value of k_{max} for the BVNS algorithm. The constructive algorithm with an α value of 0.25 and a τ value of 0.2 has been used to build the initial solutions for BVNS. Table 3 shows the obtained results with different values of k_{max}.

As it can be seen, the best value is obtained when the k_{max} value is set to 0.5. Nevertheless, the required computing time is lower when a $k_{max} = 0.4$ value is set, obtaining an average deviation value of 0.21%, which is assumable in favour of lower computing times required in the context of the OCDP.

4.2 Final Experiments

In this section, the proposed algorithm is compared against the best method found in the literature [29]. This method proposes an adaptation of the Density Peaks clustering algorithm [23] in an evolutionary context to solve the OCDP. These experiments has been performed over the whole set of 57 instances. The

Table 3. Comparison of the average results obtained by the BVNS algorithm with different values of k_{max}. Best results are highlghted with bold font.

τ	Avg	Dev. (%)	Time (s)	#Best
0.1	0.3478	0.58%	**39.57**	17
0.2	0.3523	0.43%	40.60	19
0.3	0.3672	0.33%	45.72	20
0.4	0.3696	0.21%	50.91	20
0.5	**0.3821**	**0.13%**	55.04	**24**

report table shows the results for different instances sizes, that have been divided in four groups depending on the number of nodes: from 1 to 2500, from 2500 to 5000, from 5000 to 7500 and from 7500 to 10 000 (Table 4). For the sake of fairness, both algorithms, BVNS and EADP, have been executed in the same machine. The first one has been executed with the best configuration experimentally found, and the second one with the configuration suggested by the authors.

Table 4. Comparison of Basic Variable Neighborhood Search (BVNS) and EADP configured as stated in [29] for the synthetic LFR networks.

	BVNS				EADP			
	Avg.	Dev (%)	Time (s)	#Best	Avg.	Dev (%)	Time (s)	#Best
$1 \leq n < 2500$	0.219	3.399	5.501	13	0.216	34.855	**0.507**	2
$2500 \leq n < 5000$	0.283	1.565	27.736	15	0.238	37.061	**3.243**	0
$5000 \leq n < 7500$	0.237	1.792	48.293	18	0.236	37.102	**10.781**	0
$7500 \leq n \leq 10000$	0.237	1.537	87.124	9	0.234	37.171	**36.943**	0
Average	**0.264**	**2.073**	42.164	**55**	0.231	36.547	**12.869**	2

Regarding at these results, it can be said that the BVNS proposal is slightly better in terms of the objective function values than the EADP algorithm. The EADP algorithm provides a better performance in terms of the computing time required to solve the networks under evaluation, but it only reaches the best solution in 2 out of 57 instances, which suggest that the BVNS algorithm is able to find better solutions in an average of 42.164 s, that is a reasonable computing time for the problem.

5   Conclusions and Future Work

In this paper, a new metaheuristic method for overlapping community detection in social network is proposed. It is based on Basic Variable Neighborhood Search

(BVNS) framework, using a Greedy Randomized Adaptive Search Procedure (GRASP) as constructive method for initial solutions. The modularity adapted to the context of the OCDP is used as evaluation metrics.

The performed experiments show that the proposed algorithm is able to produce high quality solutions in terms of Overlapping Community Detection Problem (OCDP), slightly better than the best method found in the literature, which is based on an adaptation of Density Peaks Clustering algorithm (EADP, Extended Adaptive Density Peaks).

In future works, it would be interesting to analyze the performance of the approach in real-world instances, as well as improve the quality of the proposal in terms of computing time and objective function values.

References

1. Camacho, D., Panizo-LLedot, A., Bello-Orgaz, G., Gonzalez-Pardo, A., Cambria, E.: The four dimensions of social network analysis: an overview of research methods, applications, and software tools. Inf. Fusion **63**, 88–120 (2020)
2. Casas-Martínez, P., Casado-Ceballos, A., Sánchez-Oro, J., Pardo, E.G.: Multiobjective grasp for maximizing diversity. Electronics **10**(11), 1232 (2021)
3. Duarte, A., Escudero, L.F., Martí, R., Mladenovic, N., Pantrigo, J.J., Sánchez-Oro, J.: Variable neighborhood search for the vertex separation problem. Comput. Oper. Res. **39**(12), 3247–3255 (2012)
4. Duarte, A., Sánchez-Oro, J., Mladenović, N., Todosijević, R.: Variable neighborhood descent. Handbook of Heuristics, pp. 341–367 (2018)
5. Feo, T.A., Resende, M.G.C.: A probabilistic heuristic for a computationally difficult set covering problem. Oper. Res. Lett. **8**(2), 67–71 (1989)
6. Feo, T.A., Resende, M.G.C., Smith, S.H.: A Greedy Randomized Adaptive Search Procedure for Maximum Independent Set. Operations Research **42**(5), 860–878 (1994)
7. Fernandez, M., Gonzalez-Pardo, A., Alani, H.: Radicalisation influence in social media. J. Web Sci. **6** (2019)
8. García-López, F., Melián-Batista, B., Moreno-Pérez, J.A., Moreno-Vega, J.M.: The parallel variable neighborhood search for the p-median problem. J. Heuristics **8**(3), 375–388 (2002)
9. Gregory, S.: Finding overlapping communities in networks by label propagation. New J. Phys. **12**(10), 103018 (2010)
10. Hansen, P., Mladenović, N., Brimberg, J., Pérez, J.A.M.: Variable neighborhood search. In: Gendreau, M., Potvin, J.-Y. (eds.) Handbook of Metaheuristics. ISORMS, vol. 272, pp. 57–97. Springer, Cham (2019). https://doi.org/10.1007/978-3-319-91086-4_3
11. Hansen, P., Mladenović, N., Perez-Britos, D.: Variable neighborhood decomposition search. J. Heuristics **7**(4), 335–350 (2001). https://doi.org/10.1023/A:1011336210885
12. Herrán, A., Colmenar, M.J., Duarte, A.: A variable neighborhood search approach for the vertex bisection problem. Inf. Sci. **476**, 1–18 (2019). https://doi.org/10.1016/J.INS.2018.09.063
13. Jaki, S., Smedt, T.D.: Right-wing german hate speech on twitter: Analysis and automatic detection. CoRR arXiv:abs/1910.07518 (2019)

14. Kumpula, J.M., Kivelä, M., Kaski, K., Saramäki, J.: Sequential algorithm for fast clique percolation. Phys. Rev. E **78**, 026109 (2008)
15. Lázár, A., Abel, D., Vicsek, T.: Modularity measure of networks with overlapping communities. EPL (Europhysics Letters) **90**(1), 18001 (2010)
16. Lozano-Osorio, I., Martínez-Gavara, A., Martí, R., Duarte, A.: Max-min dispersion with capacity and cost for a practical location problem. Expert Syst. Appl. **200**, 116899 (2022)
17. Luo, W., Zhang, D., Jiang, H., Ni, L., Hu, Y.: Local community detection with the dynamic membership function. IEEE Trans. Fuzzy Syst. **26**(5), 3136–3150 (2018)
18. Martin-Santamaria, R., Sánchez-Oro, J., Pérez-Peló, S., Duarte, A.: Strategic oscillation for the balanced minimum sum-of-squares clustering problem. Inf. Sci. **585**, 529–542 (2022)
19. Newman, M.E.J.: Modularity and community structure in networks. Proc. Natl. Acad. of Sci. **103**(23), 8577–8582 (2006)
20. Page, L., Brin, S., Motwani, R., Winograd, T.: The pagerank citation ranking: Bringing order to the web. Technical report, Stanford Info Lab (1999)
21. Pérez-Peló, S., Sánchez-Oro, J., Duarte, A.: Detecting weak points in networks using variable neighborhood search. In: Sifaleras, A., Salhi, S., Brimberg, J. (eds.) ICVNS 2018. LNCS, vol. 11328, pp. 141–151. Springer, Cham (2019). https://doi.org/10.1007/978-3-030-15843-9_12
22. Pérez-Peló, S., Sánchez-Oro, J., Gonzalez-Pardo, A., Duarte, A.: A fast variable neighborhood search approach for multi-objective community detection. Appl. Soft Comput. **112**, 107838 (2021)
23. Rodriguez, A., Laio, A.: Clustering by fast search and find of density peaks. Science **344**(6191), 1492–1496 (2014). https://doi.org/10.1126/science.1242072
24. Sánchez-Oro, J., Martínez-Gavara, A., Laguna, M., Duarte, A., Martí, R.: Variable neighborhood descent for the incremental graph drawing. Electron. Notes Discrete Math. **58**, 183–190 (2017). https://doi.org/10.1016/j.endm.2017.03.024, http://www.sciencedirect.com/science/article/pii/S1571065317300604. 4th International Conference on Variable Neighborhood Search
25. Sánchez-Oro, J., López-Sánchez, A.D., Colmenar, M.J.: A general variable neighborhood search for solving the multi-objective open vehicle routing problem. J. Heuristics 1–30 (2017). https://doi.org/10.1007/s10732-017-9363-8
26. Swani, K., Milne, G.R., Brown, B.P., Assaf, A.G., Donthu, N.: What messages to post? evaluating the popularity of social media communications in business versus consumer markets. Ind. Mark. Manag. **62**, 77–87 (2017)
27. Whang, J.J., Gleich, D.F., Dhillon, I.S.: Overlapping community detection using neighborhood-inflated seed expansion. IEEE Trans. Knowl. Data Eng. **28**(5), 1272–1284 (2016)
28. Xie, J., Kelley, S., Szymanski, B.K.: Overlapping community detection in networks: The state-of-the-art and comparative study. ACM Comput. Surv. **45**(4) (2013). https://doi.org/10.1145/2501654.2501657
29. Xu, M., Li, Y., Li, R., Zou, F., Gu, X.: EADP: an extended adaptive density peaks clustering for overlapping community detection in social networks. Neurocomputing **337**, 287–302 (2019)
30. Yang, J., Leskovec, J.: Overlapping community detection at scale: a nonnegative matrix factorization approach. In: Proceedings of the Sixth ACM International Conference on Web Search and Data Mining, p. 587–596 (2013)

A Simulation-Based Variable Neighborhood Search Approach for Optimizing Cross-Training Policies

Moustafa Abdelwanis(✉)📖, Nenad Mladenovic📖, and Andrei Sleptchenko📖

Research Center on Digital Supply Chain and Operations Management,
Department of Industrial and Systems Engineering, Khalifa University,
P.O. Box 127788, Abu Dhabi, United Arab Emirates
{100059843,nenad.mladenovic,andrei.sleptchenko}@ku.ac.ae

Abstract. We study cross-training policies in a single multi-skill, multi-server repair facility with an inventory of ready-to-use spare parts. The repair facility has an inventory facility for different spare parts. If available, the failed spare parts are immediately replaced with new ones from inventory. Otherwise, the spare parts are backordered with penalty costs. This paper proposes a model to optimize skill assignments to minimize the system's total cost, including servers, training, holding, and backorder costs. We develop a simulation-based variable neighborhood search approach, where we use discrete event simulation to evaluate backorder and holding costs under stochastic demand and service times. The simulation model is integrated with the optimization model to find the optimal skill distribution between servers. We tested the performance of our proposed framework by comparing its results with optimal solutions for small-size cases obtained using brute-force optimization. Also, we compared the performance of the proposed VNS algorithm to GA.

Keywords: Variable Neighborhood Search · Multi-skilled repair servers · Spare parts inventory · Cross-Training · Simulation-Based Optimization

1 Introduction

Cross-training is the strategy where servers are trained to do more than one task in the workplace. Training servers to do different tasks help improve the workplace's agility in various conditions, as servers will be better at handling multiple tasks and adapting new skills. Servers with various skills also help the workplace handle sudden demand surges that could lead to short-staffed departments while others are overstaffed [2]. Cross-training can also benefit the workplace in dealing with external challenges like increased competition in the global economy, difficult recessions, and technological advancements [1]. Companies might downsize their business during challenging economic crises. Multi-skilled servers can help their business weather the storm and adapt to the new business structure.

A. Sleptchenko et al. (Eds.): ICVNS 2022, LNCS 13863, pp. 42–57, 2023.
https://doi.org/10.1007/978-3-031-34500-5_4

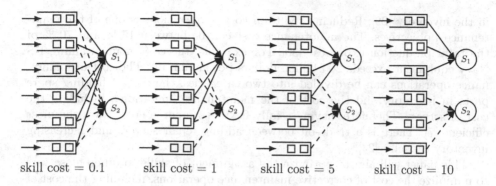

skill cost = 0.1   skill cost = 1   skill cost = 5   skill cost = 10

Fig. 1. Optimized Cross-training with different cost parameters [26].

Cross-training can be applied to human servers or machines. Training human servers can be in the form of adding skills to make them able to handle different tasks, while in the case of machines, cross-training means upgrading the software of machines to add more capabilities.

While cross-training has many benefits to the business, there are inevitable costs and drawbacks to training human servers. In addition to the psychological effects of cross-training, training human servers to do different tasks takes time, and money [2]. Moreover, there is a cost of planning how many servers to train and what skills they should have to reduce the cost or increase the system's service level.

Qin et al. in [21] listed the different levels of cross-training. The authors categorized cross-training into four levels: no cross-training, chaining, pooling, and full cross-training. No cross-training means that each worker will have no more than one dedicated task. Chaining is when each server will be assigned two tasks following each other. Pooling is when there are clusters of similar jobs, and each server in a cluster can perform all the tasks of the cluster but cannot perform any of the tasks of other clusters. In full cross-training, servers can perform any of the tasks, which is the most flexible level of cross-training. In our case, however, we are looking for fully optimized assignments, as certain cross-training patterns will be suboptimal. For example, in the particular cases presented in Fig. 1, the optimized cross-training does not follow any pattern, and as the skill costs increase, servers become more and more specialized. For more details on the experiments on optimized cross-training please refer to [26].

Cross-training servers can benefit various fields, such as cross-training maintenance servers, call center operators, and healthcare workers. However, there is a trade-off between adding skills to servers and the total cost of operations.

In this project, we focus on maintenance operations. Efficient maintenance aims to prevent the loss of value incurred to the systems during their lifetime, whether it is preventative maintenance, corrective maintenance, or condition-based maintenance [4]. One of the maintenance goals is to reduce an asset's downtime. Increased asset downtime can be due to the availability of spare parts in the inventory [29]. Reducing the total cost of operations is one of the maintenance's objectives. The maintenance cost can be between 15 % and 70 % of the total production expenses [30]. For example, the cost of maintaining US Navy equipment exceeds 200 billion dollars per annum [9]. The cost of maintenance operations can be divided into two categories: the cost of holding spare parts in inventory and the cost of the maintenance operations, including the cost of servers. Therefore, having a fully flexible system is not always the most efficient [7]. There is a trade-off between adding repair servers and increasing inventory capacity [23].

This paper will aim to optimize the assignment of skills to different servers to minimize the cost of corrective maintenance operations, including the cost of servers, training, holding costs of spare parts, and expected backorder costs. We propose a simulation-based approach to optimize skill-server assignments using Variable Neighborhood Search.

The rest of the paper is organized as follows in Sect. 2, we discuss different models in the literature related to cross-training servers. In Sect. 3, we discuss the proposed mathematical model of our system. In Sect. 4, we present our simulation-based framework. The experiments used for benchmarking and tuning our model are discussed in Sect. 5. In Sect. 6, we present the conclusion and future research areas.

2   Literature Review

This section summarizes some cross-training models in various fields like healthcare, call centers, production, and maintenance. In healthcare, Paul and MacDonald [20] developed a model to find the number of regular and cross-trained nurses that will meet the required service levels at a minimum cost. Their objective was to minimize costs, including the staffing cost (salaries) and the expected staff shortage costs, which was the cost of hiring temporary staff. They used evolutionary algorithm and search algorithm as their optimization heuristic. They analytically evaluated the service level under stochastic demand, as their model only considered two departments.

Munoz and Bastian [19] developed a discrete event simulation model to evaluate a call center's cost and capacity under different cross-training conditions. They compared the different configurations based on their cost, including the cost of training and the operators' salaries. They concluded that cross-training could help reduce the cost of service.

Agnihothri et al. [2] modeled a simple service system with multi-servers and two job types in a service workshop environment. Some of the servers are multi-skilled and can perform the two jobs. Their objective was to minimize the mean service cost and delay cost per time. They used a simulation model to investigate different parameters like the utilization and efficiency of servers.

Schober et al. [22] modeled the effect of cross-training and different qualification profiles on flow shops' quality and service levels with two production lines in a production environment. They used a multi-objective genetic algorithm-based simulation to find the optimum solution of minimizing the sum of skills and maximizing the service levels of the flow shop. Depending on the service level required, cross-training policies can be identified. Low and medium service levels can be identified with line-wise and stage-wise cross-training, as high service levels can only be reached with complete cross-training.

Altendorfer et al. [5] built on the same model and added the concept of stochastic absence of employees to the model. They found that stochastic absence reduces the service level of the flow shop and cross-training was the way to reduce the effect of the problem of stochastic absence.

Several studies in the literature concerning optimizing the assignment of skills to servers to minimize the cost of the system, including the cost of servers, cross-training, and inventory cost [3, 25–27]. Sleptchenko et al. [25] developed a simulation-based optimization framework using Particle Swarm Optimization (PSO) as the optimization heuristic. Sleptchenko et al. [26] used Genetic Algorithm (GA) instead of PSO, and extensive numerical analysis was done on real-life size problems. Cross-training was found to help reduce the cost of the system by 84%.

Al-Khatib et al. [3] studied the modular architecture of repair facilities. The authors used simulation-based optimization with PSO as their optimization heuristic. They did an experimental analysis of the finding and how each cost parameter affected their optimal solution. Turan et al. [27] modeled pooled system design instead of full cross-training to mitigate the system's risk. The authors used a discrete event simulation model integrated with improved reduced variable neighborhood search (IRVNS).

3 Problem Statement

This study focuses on a multi-server, multi-skill service workshop with full cross-training shown in Fig. 2. The repair facility has an inventory facility of different ready-to-use spare parts. In the event of a failure, the failed part is sent to the repair facility. A new one is supplied from the inventory facility to meet the demand. If the failed part is unavailable, the demand is backordered, and a new part is supplied as soon as possible.

3.1 Model Assumptions

The following assumptions were made to model the service repair facility mentioned above:

- The arrival rates of failed parts are modeled using a Poisson distribution with constant rates. This assumption is quite realistic according to previous research [26].
- The servers' repair times are mutually independent and modeled using an exponential distribution.
- The inventory holding costs are linear in the initial inventory levels (initially acquired inventory).

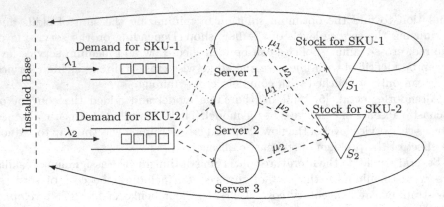

Fig. 2. Service workshop architecture for two spare parts types, three Servers, and partial flexibility [26].

- Backorder costs happen when the failed spare part is not available, and they are calculated for every time unit for every unavailable part.
- Cross-training costs happen when an operational server has two or more skills, and they are calculated based on the additional skills of the servers.
- the expected backorder costs are evaluated using steady-state probabilities on an infinite time planning horizon.

3.2   Variable Definition

Index Sets:

N: number of distinct types of repairable parts.
M: number of different servers in the repair facility.

Decision variables:

S_i: initial inventory level for repairable part of type i, $(i = 1, \ldots, N)$,
y_j: binary variable denoting that server j has assigned any skills, $(j = 1, \ldots, M)$,
x_{ij}: binary variable indicating that server j can repair any part of type i.
α_{ij}: Percentage of parts i assigned to server j $(i = 1, \ldots, N \; j = 1, \ldots, M)$

Problem parameters:

λ_i: failure rates of repairable part of type i, $(i = 1, \ldots, N)$,
μ_i: service rates of repairable part of type i, $(i = 1, \ldots, N)$,
h_i: the holding cost for repairable parts of type i per time unit $(i = 1, \ldots, N)$,
 b: the backorder penalty cost per time unit per part (e.g., downtime costs due to a lack of spare parts),
 f: the operational cost of a server per time unit (e.g., annual salary),
c_i: the cost of training servers to have the skill to repair failed part i per time unit, $(i = 1, \ldots, N)$.

3.3 Mathematical Model

In this study, the mathematical model aims to optimize skill assignment to mini-mize the total operational cost of the repair facility, including holding, backorder, servers, and cross-training costs.

$$\min_{\mathbf{S}_i, X, y_j} \left[\sum_{i=1}^{N} h_i S_i + \sum_{j=1}^{M} f y_j + \sum_{i=1}^{N} \sum_{j=1}^{M} c_i x_{ij} + b \sum_{i=1}^{N} \text{EBO}_i (\mathbf{S_i}, \mathbf{X}) \right] \qquad (1)$$

The cost terms in the objective function (1) are per unit of time. The holding, servers, and cross-training costs are linear in the decision variables S_i, X, and y_j. However, the expected backorder cost is estimated given the skill server assignment matrix X and the initial inventory level S_i. The expected backorder is a nonlinear function and cannot be evaluated analytically [25,26].

The mathematical model is subjected to sets of constraints presented below:

$$\sum_{j=1}^{M} \alpha_{ij} = 1, \qquad\qquad i = 1, \ldots, N \qquad (2)$$

$$\sum_{i=1}^{N} \alpha_{ij} \frac{\lambda_i}{\mu_i} \leq (1 - \varepsilon), \qquad j = 1, \ldots, M \qquad (3)$$

$$\alpha_{ij} \leq x_{ij}, \qquad\qquad i = 1, \ldots, N, \ j = 1, \ldots, M \qquad (4)$$

$$x_{ij} \leq y_j, \qquad\qquad i = 1, \ldots, N, \ j = 1, \ldots, M \qquad (5)$$

$$x_{ij} \in \{0, 1\}, \qquad\qquad i = 1, \ldots, N, \ j = 1, \ldots, M \qquad (6)$$

$$y_j \in \{0, 1\}, \qquad\qquad j = 1, \ldots, M \qquad (7)$$

$$\alpha_{ij} \geq 0, \qquad\qquad i = 1, \ldots, N, \ j = 1, \ldots, M \qquad (8)$$

$$S_i \in \mathbb{N}_o, \qquad\qquad i = 1, \ldots, N, \ i = 1, \ldots, N \qquad (9)$$

Constraints (Eq. 2) ensures that all failure types are assigned to servers, while Constraints (Eq. 3) ensures that the servers are not over-utilized. Constraints (Eq. 4) ensure that any server can fix only the type if they have the necessary skill.

4 A Simulation Based Variable Neighborhood Search Approach

Because of the complexity of the problem stated in [26] and the non-linearity in evaluating the expected backorder costs, a simulation-based VNS is used to optimize skill assignments. Simulation-based optimization is a technique used for solving large-scale stochastic optimization problems. Recent reviews of simu-lation optimization and its applications can be found in [6, 8, 10]. Variable neigh-borhood search has been applied previously in simulation-based optimization

frameworks in [11,12]. Mladenovic and Hansen first introduced Variable Neighborhood Search (VNS) in 1997 [18]. VNS has many variants and applications
presented in [13–15]; however, basic VNS is used in this study, and it is modeled
as in Algorithm 1. A flow diagram of the simulation-based VNS approach is
shown in Fig. 3.

Algorithm 1: BASIC VNS ALGORITHM

    Result: $BasicVNS$ (Kmax, x, N)
    $x \leftarrow InitialSolution$;
    $K \leftarrow 1$;
    for $K \leq Kmax$ **do**
        | $y \leftarrow shake(x, a, b, K)$;
        | $y' \leftarrow localsearch(y, a, b)$;
        | **if** y' *is better than* x **then**
        | | $x \leftarrow y'; K \leftarrow 1$;
        | **end**
    end
    return x;

4.1   Initial Solution

The first step of Basic VNS is to generate a feasible initial solution. In our study,
we generate two starting initial solutions. The first solution is obtained by solving
an LP problem that minimizes the number of skills in the skill assignment matrix
with objective function (10). On the other hand, the second solution is obtained
by solving an LP problem that minimizes the number of servers using objective
function (11). Both LP solutions must satisfy constraints (2-8). We then evaluate
the total cost of the generated two solutions, and the solution with the lowest
cost is our VNS framework's starting solution.

$$\min_{\mathbf{X}} \sum_{i=1}^{N} \sum_{j=1}^{M} x_{ij} \tag{10}$$

$$\min_{\mathbf{X}} \sum_{j=1}^{M} y_{ij} \tag{11}$$

4.2   Local Search

In the local search step, we explore the feasible solutions given the neighborhood
structure to improve our incumbent solution. In our framework, we use hamming
distance as our neighborhood structure. In other words, we change one element in
the assignment matrix from zero to one or vice versa (To add or remove a skill

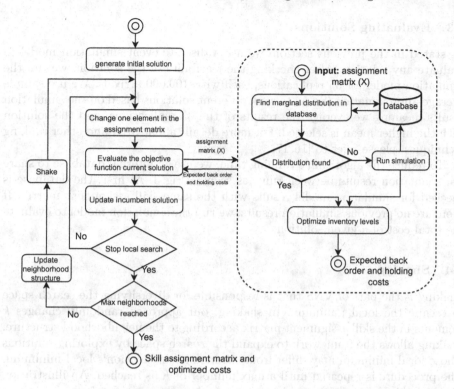

Fig. 3. Simulation-based VNS framework for optimizing skill server assignment.

from a server). Each solution's objective function is evaluated by calculating the holding, server, and cross-training costs. The expected backorder cost is evaluated through a simulation model as seen in Fig. 3. The local search step aims to find a solution with the best objective function value of feasible solutions in the defined neighborhood. An illustrative example of a local search using hamming distance is shown in Fig. 4.

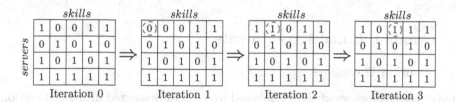

Fig. 4. Local search illustration using hamming distance.

4.3   Evaluating Solutions

As stated in the previous section, we use a discrete event simulation model to evaluate any solution. After checking the feasibility of the solution, we run the simulation model for 35 replications, each with 100,000 arrivals. We use sample average approximation (SAA) to rank different solutions based on the simulation results meaning we average the results of the 35 replications, and the solution with the higher mean is selected. For more details about SAA and other ranking techniques, please refer to [16,17].

To reduce the algorithm's running time, we use an internal database to share the simulation results between runs, as seen in Fig. 3. At first, the database is checked for simulation model results with the same skill assignment matrix. If there are no previous simulation results, we run the simulation model to evaluate the total cost of a given solution.

4.4   Shaking

Shaking is the part of VNS that is responsible for diversifying the search space to escape the local minimum. In shaking, our approach randomly changes k elements in the skill assignment matrix according to the neighborhood structure. Shaking allows the framework to expand the search space by exploring solutions whose local minimum may differ from the incumbent solution's local minimum. The procedure is repeated until a max number of K is reached. An illustration of the shaking procedure used in our approach is shown in Fig. 5.

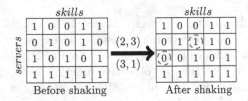

Fig. 5. Example of shaking when $k = 2$.

5   Computational Experiments

5.1   Tuning Data Sets

One of the advantages of the VNS-based framework presented in the last section is that it has only one parameter to tune, the maximum number of neighborhoods ($Kmax$). To find the value for $Kmax$ that will give us better results, we use the dataset in [24]. We use the dataset to compare the performance of our framework with different $Kmax$ values. In these experiments, we use only small cases where the optimal solution can be found using total enumeration techniques. In other words, for the small cases, the number of possible skill assignments was limited to 3000. Table 1 shows the number of feasible skill assignments for the selected cases.

Table 1. Number of feasible skill assignments (shaded in grey) for different numbers of skills and servers in the small cases.

	Number of Skills				
Number of Servers	2	3	4	5	6
2	4	13	38	117	356
3	12	62	404	-	-
4	17	165	-	-	-
5	24	446	-	-	-
6	46	1193	-	-	-

For the tuning experiments, we changed the value of $Kmax$ between 6, 8, 10, and 12. We also changed some cost parameters in the experiments, like the minimum holding cost and the cost of servers. The minimum holding cost has a value of either 1 or 100, and the cost of a server is either 10000 or 100000. Changing the cost parameters and the value of $Kmax$ results in 16 experiments, and we have 840 small cases. The total number of experiments is $16 \times 840 = 13440$ experiments. Table 2 shows the design factors and levels of the experiments.

Table 2. Design factors for tuning DOE.

Factors	Levels
Maximum number of neighborhoods (Kmax)	[6, 8, 10,12]
Machine costs	[10000, 100000]
Minimum holding cost	[1, 100]
Total Experiments	$16 \times 840 = 13440$

5.2   Tuning Results

After comparing the results of the VNS framework with the brute force procedure results, the model's average error was (0.0107 %) in the 13440 experiments. The average number of iterations was (303.4) per experiment. The model has two cases where the error was more than (2.5%); the VNS objective function values were 2.51% and 2.69% more than the optimal solution. A histogram of the relative error for different values of $Kmax$ is shown in Fig. 6. Figure 6 shows that the relative error decreases with the increased number of $Kmax$. The average error, iterations, and the number of cases where the optimal solution was found are shown in Table 3.

Table 3. Performance of VNS algorithm using different $Kmax$ values.

Kmax	Average Error	Max. Error	Average iterations	Optimum was found
6	0.021%	2.69%	200.51	3142 (93.5 %)
8	0.01%	2.51%	269.39	3223 (95.9%)
10	4.52×10^{-3} %	1.01%	338.39	3268 (97.2 %)
12	6.62×10^{-3} %	2.2%	405.43	3278 (97.5 %)

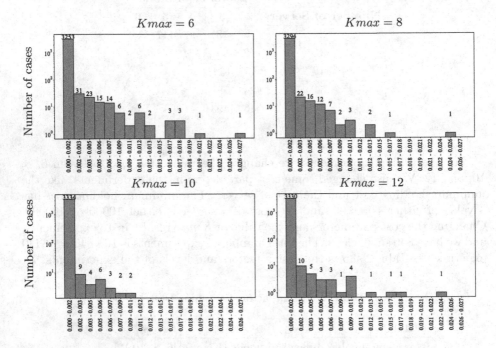

Fig. 6. Histograms of relative error for each value of $Kmax$ in logarithmic scale.

We also analyzed the model's speed for each value of $Kmax$. In the box plot shown in Fig. 7, we can see that increase in $Kmax$ would increase the number of iterations. Another point of interest to the study was how fast the model converges to the optimal solution. Figure 8 shows how fast the model reaches the optimal solution. The figure shows the number of iterations at which each model run reaches a certain percentage of error from the optimal solution.

To study the effect of our initial solution, we repeated the same experiments twice with two frameworks where the initial solutions are obtained by solving an LP with (10) and (11) as their objective functions. The three models had similar relative errors and number of iterations. However, the convergence of our model

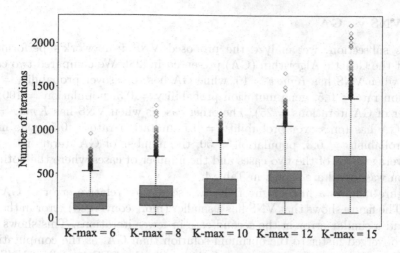

Fig. 7. Number of iterations of different cases for each value of $Kmax$.

was better than the other two models. Our model had 75% of the cases converge to 1% error in less than 100 iterations, and 75% of the cases converge to 0.1% error in less than 200 iterations. Figure 8 shows how fast our model converges to the optimal solution compared with the other two models.

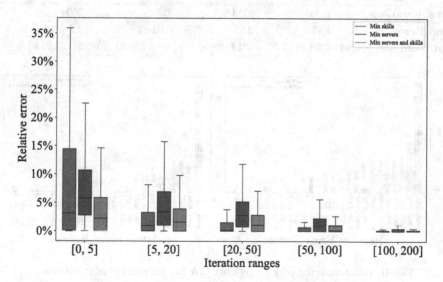

Fig. 8. Comparison of the convergences of different models using different initial solutions.

5.3   VNS vs. GA

In this subsection, we analyze the proposed VNS framework's performance against the Genetic Algorithm (GA) presented in [28]. We compared two cases; one is when VNS has $Kmax = 15$, while GA has (crossover probability $= 0.7$, mutation rate $= 0.5$, gene mutation probability $= 0.3$, population $= 100$, the number of GA iterations $= 25$). The other case is when VNS has $Kmax = 10$, while GA has (crossover probability $= 0.7$, mutation rate $= 0.5$, gene mutation probability $= 0.3$, population $= 50$, the number of GA iterations $= 25$). The average error of the two cases and the number of cases where the optimum solution was found are shown in Table 4.

Figure 9 shows a histogram of the two scenarios' relative error of GA and VNS. The figure shows that VNS has a smaller (more converged) error in the first case and a similar error in the second case in fewer iterations. This shows that VNS converged faster to the optimum solution than GA, as the computational time of one iteration (evaluating a solution using the DES model) of VNS and GA was the same.

Table 4. Comparison of the VNS framework to GA's.

	VNS		GA	
	Kmax = 10	Kmax = 15	50 populations, 25 generations	100 populations, 25 generations
Max. iterations	1188	2214	1250	2500
Avg. Error	4.52×10^{-3} %	2.52×10^{-3} %	0.01 %	3.2×10^{-3} %
Optimum was found	3278 (97.5 %)	3314 (98.63 %)	3189 (94.91 %)	3312 (98.57 %)

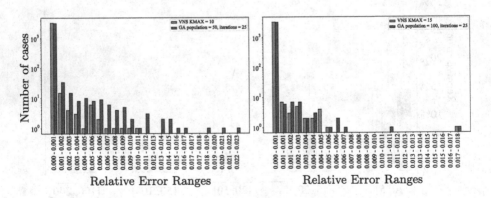

Fig. 9. Relative error of VNS against GA for different configurations.

6 Conclusion and Future Work

This paper proposed a simulation-based Variable Neighborhood Search framework for optimizing skill-server assignment in a service repair facility with full cross-training. The framework's objective is to minimize the system's total cost under stochastic failure rates and service times, including the cost of servers, cross-training, holding, and backorder costs. The proposed framework was tested using the Design of Experiments to measure its performance under different cost factors and $Kmax$ values. The experiments used small cases that can be solved using brute force procedures. The model showed promising results with an average error of (0.0107%) of the optimal solution and an average number of iterations of (303.4). The performance of the proposed algorithm was compared to GA and was found to have a less relative error in fewer iterations.

For future studies, we want to extend the work and test our proposed framework for large cases or realistic-size problems. We also want to investigate the effect of different prioritization policies on the system.

Acknowledgements. This research is supported by Khalifa University under Grant No. FSU-2020-19 and Award No. RC2 DSO.

References

1. Abrams, C., Berge, Z.: Workforce cross training: a re-emerging trend in tough times. J. Workplace Learn. **22**(8), 522–529 (2010). https://doi.org/10.1108/13665621011082882
2. Agnihothri, S.R., Mishra, A.K., Simmons, D.E.: Workforce cross-training decisions in field service systems with two job types. J. Oper. Res. Soc. **54**(4), 410–418 (2003). https://doi.org/10.1057/palgrave.jors.2601535
3. Al-Khatib, M., Turan, H.H., Sleptchenko, A.: Optimal skill assignment with modular architecture in spare parts supply systems. 2017 4th International Conference on Industrial Engineering and Applications, ICIEA 2017, pp. 136–140 (2017). https://doi.org/10.1109/IEA.2017.7939194
4. Alrabghi, A., Tiwari, A.: State of the art in simulation-based optimisation for maintenance systems. Comput. Ind. Eng. **82**, 167–182 (2015). https://doi.org/10.1016/j.cie.2014.12.022
5. Altendorfer, K., Schober, A., Karder, J., Beham, A.: Service level improvement due to worker cross training with stochastic worker absence. Int. J. Prod. Res. (2020). https://doi.org/10.1080/00207543.2020.1764126
6. Amaran, S., Sahinidis, N., Sharda, B., Bury, S.: Simulation optimization: a review of algorithms and applications. Ann. Oper. Res. **240** (2016). https://doi.org/10.1007/s10479-015-2019-x
7. Bassamboo, A., Randhawa, R., Van Mieghem, J.: A little flexibility is all you need: on the asymptotic value of flexible capacity in parallel queuing systems. Oper. Res. **60**, 1423–1435 (2012). https://doi.org/10.2307/23323709
8. Chica, M., Juan, A.A., Bayliss, C., Cord, O., Kelton, W.D.: Why simheuristics? Benefits, limitations, and best practices when combining metaheuristics with simulation. Sort-Stat. Oper. Res. Trans. **44**(2), 0311–0334 (2020). https://doi.org/10.2436/20.8080.02.104

9. Fang, L., Zhaodong, H.: System dynamics based simulation approach on corrective maintenance cost of aviation equipments. Procedia Eng. **99**, 150–155 (2015). https://doi.org/10.1016/j.proeng.2014.12.519

10. Gosavi, A.: Simulation-Based Optimization. ORSIS, vol. 55. Springer, Boston (2015). https://doi.org/10.1007/978-1-4899-7491-4

11. Gruler, A., Panadero, J., Armas, J., Pérez, J., Juan, A.: A variable neighborhood search simheuristic for the multiperiod inventory routing problem with stochastic demands. Int. Trans. Oper. Res. **27** (2018). https://doi.org/10.1111/itor.12540

12. Gruler, A., Panadero, J., de Armas, J., Moreno Pérez, J.A., Juan, A.A.: Combining variable neighborhood search with simulation for the inventory routing problem with stochastic demands and stock-outs. Comput. Ind. Eng. **123**, 278–288 (2018). https://doi.org/10.1016/j.cie.2018.06.036

13. Hansen, P., Mladenović, N.: Variable neighborhood search. Search Methodologies: Introductory Tutorials in Optimization and Decision Support Techniques pp. 211–238 (2005)

14. Hansen, P., Mladenovic, N., Moreno-Pérez, J.: Variable neighbourhood search: methods and applications. 4OR **175**, 367–407 (2010). https://doi.org/10.1007/s10479-009-0657-6

15. Hansen, P., Mladenovic, N., Todosijević, R., Hanafi, S.: Variable neighborhood search: basics and variants. EURO J. Comput. Optim. **5** (2016). https://doi.org/10.1007/s13675-016-0075-x

16. Kim, S.H., Nelson, B.L.: Recent advances in ranking and selection. In: 2007 Winter Simulation Conference, pp. 162–172 (2007). https://doi.org/10.1109/WSC.2007.4419598

17. Kleywegt, A.J., Shapiro, A., Homem-de Mello, T.: The sample average approximation method for stochastic discrete optimization. SIAM J. Optim. **12**(2), 479–502 (2002). https://doi.org/10.1137/S1052623499363220

18. Mladenović, N., Hansen, P.: Variable neighborhood search. Comput. Oper. Res. **24**(11), 1097–1100 (1997). https://doi.org/10.1016/S0305-0548(97)00031-2

19. Munoz, D.A., Bastian, N.D.: Estimating cross-training call center capacity through simulation. J. Syst. Sci. Syst. Eng. **25**(4), 448–468 (2016). https://doi.org/10.1007/s11518-015-5286-9

20. Paul, J.A., MacDonald, L.: Modeling the benefits of cross-training to address the nursing shortage. Int. J. Prod. Econ. **150**, 83–95 (2014). https://doi.org/10.1016/j.ijpe.2013.11.025

21. Qin, R., Nembhard, D.A., Barnes, W.L.: Workforce flexibility in operations management. Surv. Oper. Res. Manage. Sci. **20**(1), 19–33 (2015). https://doi.org/10.1016/j.sorms.2015.04.001

22. Schober, A., Altendorfer, K., Karder, J., Beham, A.: Influence of workforce qualification on service level in a flow shop with two lines. IFAC-PapersOnLine **52**(13), 553–558 (2019). https://doi.org/10.1016/j.ifacol.2019.11.216

23. Sleptchenko, A., Van der Heijden, M.C., Van Harten, A.: Trade-off between inventory and repair capacity in spare part networks. J. Oper. Res. Soc. **54**(3), 263–272 (2003). https://doi.org/10.1057/palgrave.jors.2601511

24. Sleptchenko, A.: Simulation results for multi-class multi-server queueing systems with cross-training (2017). https://doi.org/10.17632/48fctynd77.2

25. Sleptchenko, A., Elmekkawy, T., Turan, H.H., Pokharel, S.: Simulation based particle swarm optimization of cross-training policies in spare parts supply systems. In: 9th International Conference on Advanced Computational Intelligence, ICACI 2017, pp. 60–65. Institute of Electrical and Electronics Engineers Inc, July 2017. https://doi.org/10.1109/ICACI.2017.7974486

26. Sleptchenko, A., Turan, H.H., Pokharel, S., ElMekkawy, T.Y.: Cross-training policies for repair shops with spare part inventories. Int. J. Prod. Econ. **209**(August
2017), 334–345 (2019). https://doi.org/10.1016/j.ijpe.2017.12.018
27. Turan, H.H., Sleptchenko, A., Pokharel, S., ElMekkawy, T.Y.: A sorting based
efficient heuristic for pooled repair shop designs. Comput. Oper. Res. **117** (2020).
https://doi.org/10.1016/j.cor.2020.104887
28. Turan, H.H., Pokharel, S., Sleptchenko, A., Elmekkawy, T.Y.: Integrated optimization for stock levels and cross-training schemes with simulation-based genetic algorithm. In: 2016 International Conference on Computational Science and Computational Intelligence (CSCI), pp. 1158–1163 (2016). https://doi.org/10.1109/CSCI.
2016.0219
29. Van Horenbeek, A., Buré, J., Cattrysse, D., Pintelon, L., Vansteenwegen, P.: Joint
maintenance and inventory optimization systems: a review. Int. J. Prod. Econ.
143(2), 499–508 (2013). https://doi.org/10.1016/j.ijpe.2012.04.001
30. Wang, L., Chu, J., Mao, W.: An optimum condition-based replacement and
spare provisioning policy based on Markov chains (2008). https://doi.org/10.1108/
13552510810909984

Multi-objective Variable Neighborhood Search for Improving Software Modularity

Javier Yuste[✉][iD], Eduardo G. Pardo[iD], and Abraham Duarte[iD]

Universidad Rey Juan Carlos, Móstoles, 28933 Madrid, Spain
{javier.yuste,eduardo.pardo,abraham.duarte}@urjc.es

Abstract. Search-Based Software Engineering is a research area that aims to tackle software engineering tasks as optimization problems. Among the problems in this area, we can find the Software Module Clustering Problem (SMCP). This problem, which has been proved to be $\mathcal{NP}$-hard, focuses on finding the best organization of a software project in terms of modularity. Since modular code is easier to understand, the result is an increase of the quality of software projects and thus a reduction of the costs associated to their maintenance. To tackle the SMCP, software projects are often modeled as graphs that represent the dependencies between different components. In this work, we study two well-known multi-objective approaches for the SMCP: the Maximizing Cluster Approach (MCA) and the Equal-size Cluster Approach (ECA). Each of these variants is composed of 5 different objectives. We propose a heuristic algorithm based on the Multi-Objective Variable Neighborhood Descent (MO-VND) schema to tackle the aforementioned variants and we introduce three neighborhoods to be explored within the algorithm. Finally, we compare the performance of our proposal with the performance of NSGA-III over a dataset of real software projects. The results show that the proposed algorithm is competitive when tackling the MCA, and some ideas are given to increase its efficiency when tackling the ECA.

Keywords: Software Module Clustering · Search-Based Software Engineering · Modularization Quality · Heuristics · Multi-Objective Optimization

1 Introduction

Software systems are developed and maintained following a well-known Software Development Life-Cycle, which contains several activities. These activities are performed in different phases, such as planning, analysis, design, implementation, testing, operation, and maintenance. Among them, the most costly phase is maintenance, which is frequently responsible for up to 75 % of the total costs [4]. Interestingly, most efforts in this phase are dedicated to comprehending the existent software. Without a proper understanding of the code base, its maintenance

A. Sleptchenko et al. (Eds.): ICVNS 2022, LNCS 13863, pp. 58–68, 2023.
https://doi.org/10.1007/978-3-031-34500-5_5

is costly and prone to error. To ease this task, the code is usually divided into several components, facilitating the understanding of each one in a separate way. Moreover, these components are then organized in different modules to ease the understanding of each module. This organization follows the principle of modularization, trying to maximize the *"software attributes that provide a structure of highly independent components"* [11]. That is, a structure is said to be modular when its components are highly independent of others. This is, components in the same module are strongly related (high cohesion), and components in different modules are loosely connected (low coupling). The concept of a software component can be understood in different ways (e.g., a class, a file, a folder, etc.) and it is also frequent to see that the terms "module" and "component" are used interchangeably [11]. For the sake of understandability, in this work we will use the term "component" for atomic units (e.g., classes and source code files) and the term "module" for groups of components (e.g., folders and packages).

Search-Based Software Engineering is a research area that focuses on tackling software engineering problems as optimization problems [6]. In this area, the Software Module Clustering Problem (SMCP) is an optimization problem which objective is to maximize the modularity of software systems. Although the overall objective of the SMCP is clear, the simple dogma of increasing cohesion and decreasing coupling does not seem to work [22]. If it was taken to the extreme, the best organization for any software project would have a single module containing all components. As such a trivial solution is not useful for easing maintainability, other variables, like the number of modules, must also be considered. Accordingly, one of the first and most used objective functions in the literature, Modularization Quality (MQ), computes the value of the modularity as a tradeoff between coupling and cohesion while implicitly considering the number of modules [14]. Two variants of this metric, BasicMQ and TurboMQ, were later proposed [16, 17]. These metrics have been widely used in the literature [10, 25, 28] and have served as the basis for alternative objective funtions [21]. However, by learning from previous work in the SMCP, some researchers have come to realize that software engineers usually organize their projects considering different context variables in a subjective way, which results in a process that is not necessarily systematic and repeatable [22]. Therefore, a multi-objective approach seems more pertinent for the SMCP than a monoobjective one since: i) different conflicting objectives might represent the modularity of software systems better than a single metric; and ii) presenting a set of good solutions according to different metrics allows end users to introduce their subjective experience in the process, by deciding on the best one from their viewpoint. Among the different multi-objective approaches proposed for the SMCP [18, 24], perhaps the most popular are two of the first ones introduced in the literature: the Maximizing Cluster Approach (MCA) and the Equal-size Cluster Approach (ECA) [26]. These approaches, which combined five different objectives each, were later widely used and extended [1, 2, 5, 27]. Interestingly, the MQ metric was included as one of the five objectives in both approaches.

Regardless of the objective functions considered, the combinatorial nature of the problem makes it impracticable to evaluate all possible solutions for most software projects. Indeed, the SMCP has been proven to be $\mathcal{NP}$-hard [3]. As a consequence, exact methods are not suitable for this problem. Instead, approximate search-based methods are needed to tackle the SMCP [9,29]. Historically, evolutionary approaches, which are a subset of bioinspired search-based algorithms, have been remarkably popular in the literature to solve the SMCP [29]. However, some works have recently proposed the use of trajectory-based algorithms, which progressively improve a given solution through small modifications, until a local optimum is found, achieving better results than previous evolutionary algorithms [20,23,30]. In the context of multi-objective optimization, evolutionary approaches have been traditionally favored, since they naturally work with a set (population) of solutions. Following this trend, some authors have proposed approaches to tackle the SMCP based on genetic algorithms [12], Artificial Bee Colony methods [5], or well-known evolutionary algorithms such as PESA [1] or NSGA-III [18].

In this paper, we propose a Multi-Objective Variable Neighborhood Descent (MO-VND) method to tackle the SMCP problem in a multi-objective context, using a trajectory-based metaheuristic. In particular, we study the Maximizing Cluster Approach and the Equal-size Cluster Approach sets of objectives. The MO-VND algorithm proposed explores three different neighborhoods to obtain high-quality solutions in short computing times. We then compare the results obtained by our proposal with those obtained by the well-known Non-dominated Sorting Genetic Algorithm III (NSGA-III) over a dataset of 14 real world software projects. The proposed method is shown to be competitive, and some improvements are highlighted based on the obtained results.

2   Problem Definition

In the SMCP, the software projects are often modeled in a graph structure, where components are represented as vertices, dependencies between components are represented as edges, and modules are represented as clusters or groups of vertices. More formally, a software can be modeled as an undirected weighted graph $G = (V, E, W)$, where V is the set of vertices, E is the set of edges, and W is the set of weights associated to the edges in E. In related literature, this structure is commonly named a Module Dependency Graph (MDG). Then, a solution for the SMCP can be defined as a partition of the MDG, which represents the software architecture. That is, a set M of disjoint subsets of V, such that $M = \{m_1, m_2, ..., m_k\}$, where k represents the number of modules $(1 \leq k \leq |V|)$ and each m_i, with $1 \leq i \leq k$, is a disjoint subset of V. As it can be seen, there exist two trivial solutions: i) a solution in which each vertex forms an isolated module $(k = |V|)$ and ii) a solution in which every vertex is assigned to the same module $(k = 1)$.

To evaluate the quality of the solutions, we study the SMCP based on different conflicting objectives:

1. Maximize cohesion.
2. Minimize coupling.
3. Maximize the number of modules.
4. Maximize the Modularization Quality (MQ).
5. Minimize the number of isolated modules.
6. Minimize the difference between the maximum and minimum number of vertices in a module.

The cohesion of an architecture (first objective) is calculated as the sum of the weights of edges that connect vertices belonging to the same module. In contrast, the coupling (second objective) of the solution is calculated as the sum of the weights of edges that connect vertices belonging to different modules. The number of modules (third objective) is calculated as the count of groups of vertices (k). For each module, its size corresponds to the number of vertices that belong to that module. Therefore, a module is said to be isolated (fifth objective) if it contains only one vertex and the difference in size between two modules (sixth objective) is equal to the difference in the number of vertices that belong to each module. Finally, to evaluate the MQ (fourth objective), we must calculate the Modularization Factor (MF_i) of each module m_i. Specifically, given a solution x:

$$MF_i(x) = \begin{cases} 0, & \text{if } \mu_i = 0 \\ \frac{2\mu_i}{2\mu_i + \varepsilon_i}, & \text{if } \mu_i > 0 \end{cases} \tag{1}$$

where μ_i and ε_i are the cohesion and coupling values of module m_i, respectively. Then, the value of the objective function MQ is calculated as the sum of the MF_i of each module m_i. In mathematical terms:

$$MQ(x) = \sum_{i=1}^{k} MF_i. \tag{2}$$

This calculation of the MQ metric is known as TurboMQ [17], which is a variant of the original MQ metric proposed in [14]. For a more in-depth explanation of this metric, we refer the reader to related works [30].

The aforementioned objectives were proposed in [26] to form two distinct sets of objectives: MCA and ECA. Both approaches have a set of five conflicting objectives. In particular, the MCA approach is formed by objectives 1, 2, 3, 4, and 5, whereas the ECA approach is formed by objectives 1, 2, 3, 4, and 6. As it can be noticed, the difference between both approaches is located in the fifth objective: MCA proposes the minimization of the number of isolated clusters, while ECA proposes the minimization of the difference in size between the largest and the smallest modules.

3 Algorithmic Proposal

To tackle the SMCP, we propose an algorithm based on the Variable Neighborhood Search (VNS) methodology, a general framework to solve hard combinatorial and global optimization problems [19]. The main idea behind this methodology is to systematically explore different neighborhood structures until a local optimum within all the explored neighborhoods is found. Although VNS was originally designed for mono-objective problems, it was later extended by A. Duarte et al. to tackle optimization problems in multi-objective contexts [7]. In the Multi-Objective VNS (MO-VNS) methodology, the term solution is used to denote a Pareto front, whereas the elements of the Pareto front are denoted as efficient points. This terminology allows the authors to naturally extend the original methodology for multi-objective contexts, since an improvement is now considered as the inclusion of a new efficient point in the solution. Among the variants proposed, we select to implement here an algorithm based on the Multi-Objective Variable Neighborhood Descent (MO-VND) schema described in [7]. This variant explores both the objectives and the neighborhoods given in a systematic order.

The proposed algorithm needs to generate a set of initial efficient points. These efficient points will be later improved by the MO-VND method. In this case, we propose to generate the initial solutions by using a random constructive procedure. This method starts by creating a set of $|V|$ modules, where V is the set of vertices in the current instance. Then, every vertex is inserted into an existent module. This operation is performed randomly, meaning that every vertex has an equal probability to be inserted in each of the existing modules ($1/|V|$). Once all vertices have been allocated in a module, the remaining empty modules are removed. The resulting efficient point is included in the initial solution. We repeat this process 100 times to construct the initial set of efficient points.

Once an initial solution (i.e., a set of efficient points) has been constructed, the MO-VND method is responsible for improving the set of efficient points according to the objective set tackled. The pseudocode of MO-VND is shown in Algorithm 1. This method receives three parameters: a set of efficient points E, a number of neighborhoods k_{max}, and a number of objectives r. First, a set of exploited points S_i is initialized for each objetive (step 2) and variable i is set to 1 (step 3). Then, the algorithm tries to improve the solution E while i is less than r (step 4). In each iteration, each non-exploited efficient point $x \in E \setminus S_i$ is improved within a VND-i method (steps 5–9). Notice that the set of efficient points explored in the VND-i procedure is added to S_i at each iteration in order to avoid exploring efficient points that have already been exploited (step 8). If an improvement has been made in the previous stages, then i is reset to 1 and solution E is updated to include the new efficient points generated (steps 10–12). Otherwise, the value of i is incremented by one (step 14). Finally, once all the efficient points in the solution E have been unsuccessfully explored in regards to each objective i, the algorithm returns the solution E (step 17).

In this schema, a different Variable Neighborhood Descent (VND-i) method is used to improve each objective separately. However, in contrast to the original design of VNS where we only needed to check if the final point obtained improved upon the incumbent point, in this context every point visited in the search must be checked for its possible inclusion in the solution. The pseudocode of VND-i is shown in Algorithm 2. This method receives two parameters: an efficient point x

---

Algorithm 1. Pseudocode of MO-VND

1: **procedure** MO-VND(E,k_{max},r)
2:     $S_1 \leftarrow \emptyset$, $S_2 \leftarrow \emptyset$, ..., $S_r \leftarrow \emptyset$
3:     $i \leftarrow 1$
4:     **while** $i < r$ **do**
5:         **while** $|E \setminus S_i| > 0$ **do**
6:             $x \leftarrow$ SELECT($E \setminus S_i$)   ▷ Random selection among the non-exploited points
7:                 $E_i \leftarrow$ VND-$i(x, k_{max})$
8:                 $S_i \leftarrow S_i \cup E_i$
9:         **end while**
10:        **if** MO-IMPROVEMENT(E, E') **then**
11:            $i \leftarrow 1$
12:            $E \leftarrow$ UPDATE(E, E')
13:        **else**
14:            $i \leftarrow i+1$
15:        **end if**
16:    **end while**
17:    **return** E
18: **end procedure**

---

and the maximum number of neighborhoods k_{max}. First, the variable k is set to 1 (step 2) and a set of efficient points E is initialized to contain the incumbent efficient point x (step 3). Then, while the value of k is less than k_{max} (step 4), the efficient point x is improved (step 5) and the result is added to the set E (step 6). If the resulting efficient point x' is not better than the previous solution x (step 10), then k is incremented (step 11). Otherwise, the new solution x' is saved as the current solution x (step 8) and k is reset to 1 in order to restart the exploration of all the neighborhoods (step 9). Once all neighborhoods have been unsuccessfully explored for a particular solution, the procedure returns the constructed set of efficient points E (step 14).

For the VND-i component of the MO-VND schema, we propose three different neighborhoods to be explored:

- The first neighborhood proposed (N_1) is defined by an insertion operator. This operator removes a vertex from its current module and inserts it into a different module.
- The second neighborhood (N_2) is defined by a destruction operator. This operator removes one module from the efficient point and reinserts the affected vertices, one by one, into other modules.
- The third neighborhood (N_3) is defined by an extraction operator. This operator selects some vertices from one or more modules and creates a new module with the selected vertices.

Algorithm 2. Pseudocode of VND-i

1: **procedure** VND-$i(x,k_{max})$
2:     $k \leftarrow 1$
3:     $E \leftarrow \{x\}$
4:     **while** $k < k'_{max}$ **do**
5:         $x' \leftarrow \underset{y \in N_k(x)}{\text{argmin}}\, z_i(y)$ ▷ Find the best neighbor in $N_k(x)$
6:         UPDATE$(E, N_k(x))$
7:         **if** $z_i(x') < z_i(x)$ **then**
8:             $x \leftarrow x'$
9:             $k \leftarrow 1$
10:        **else**
11:            $k \leftarrow k + 1$
12:        **end if**
13:    **end while**
14:    **return** E
15: **end procedure**

---

4   Experimental Results

In this section, we present the experimental results obtained in this research. In Sect. 4.1, we describe the set of instances used in the experimentation. In Sect. 4.2, we present the parametrization of the proposed algorithm. Finally, in Sect. 4.3, a comparison of the proposed algorithm and NSGA-III is presented.

4.1   Dataset

For the experiments described in this section, we use a set of 14 real software systems introduced by the community in [20]. Varying in size, these instances range from a minimum of 14 vertices and 20 edges to a maximum of 626 vertices and 2421 edges, with an average of 166 vertices and 876.79 edges. Particularly, the instances contained in this dataset are the following: apache_ant_taskdef, gae_plugin_core, javacc, joe, jscatterplot, jtreeview, jxlsreader, lwjgl-2.8.4, mod_ssl, net-tools, nmh, regexp, star, wu-ftpd-1.

4.2   Order of the Neighborhoods

The VND-i method explores the given neighborhoods in a systematic way. After exploring each neighborhood, the method returns to the first neighborhood if an improvement has been made in the efficient point. Else, the method continues to explore the next neighborhood. This process is repeated until all neighborhoods have been explored without improving the incumbent efficient point. Therefore, the order in which the neighborhoods are explored is important, since the first neighborhood is frequently explored more times than the following. In this experiment, we use the irace software package [13], designed to automatically configure the parameters of an algorithm. Specifically, we compare the performance of the algorithm, when exploring the neighborhoods in different orders, in terms of hypervolume and CPU time. After running the described experiment, irace reports that the best configuration found explores the neighborhoods in the following order: N_2, N_3, and N_1.

4.3   Comparison with State-of-the-Art Algorithms

In this section, we present a comparison of the experimental results obtained with the proposed MO-VND and NSGA-III. We compared both algorithms over a preliminary dataset of 14 real instances made publicly available by the community [20]. All experiments were run on an Intel Xeon Processor (Cascadelake), with 64 cores and 124 GB RAM. The Operating System used was Ubuntu 22.04 LTS. Our proposal was coded in Java 17.0.5 and using the Metaheuristic Optimization framewoRK (MORK) project v0.12 [15]. For the comparison, we used the NSGA-III implementation publicly available in jMetal v5.11 [8].

When comparing algorithms in a multi-objective context, there exist several metrics to compare the sets of efficient points generated by the methods under evaluation. Most of these metrics, additionally, need to know a reference front to evaluate the efficiency of the solutions. However, the reference front, which ideally is the set of efficient points that cannot be dominated by any feasible solution in the current objective space, is often unknown. In this case, since the reference front for each instance is unknown, we construct an approximate reference front as the union of the reference fronts generated by the algorithms under evaluation. That is, for each instance, we consider the reference front to be the union of the efficient points generated by both the MO-VND and NSGA-III methods.

Regarding the metrics used in this work, we focus our attention on Hypervolume, Coverage, Inverted Generational Distance Plus, and Generalized Spread. First, we analyze the Hypervolume (HV) of the generated fronts, which evaluates the volume of the given set in the objective space. The larger its value, the better. Second, we report the Coverage (Cov.) of each set, which indicates the number of efficient points in the solution that are dominated by the reference front. The smaller the value of Coverage, the better the solution. In third place, we consider the Inverted Generational Distance Plus (IGD+) metric, which measures the distance between a given set of efficient points and a reference front. In the case of IGD+, the smaller its value, the better the solution. Last but not least, we consider the Generalized Spread (G. Spread) metric, which measures the range of values covered by the efficient points in a given solution (i.e., the diversity of the solution). The smaller the spread, the better the solution.

In Table 1, we present the results obtained by MO-VND and NSGA-III over the set of objectives described in MCA. For each method, we report the average Hypervolume (Avg. HV), the average Coverage (Avg. Cov.), the average IGD+ (Avg. IGD+), and the average G. Spread (Avg. Spread) of the generated solutions. Additionally, we report the average CPU time consumed by each method to generate the solutions in seconds (CPUt (s)). As it can be observed,

Table 1. Comparison of the results with the algorithm presented in this work (MO-VND) and NSGA-III over the objective set proposed in MCA.

Method	Avg. HV	Avg. Cov	Avg. IGD+	Avg. G. Spread	Avg. CPUt (s)
MO-VND	**0.0877**	0.3589	0.1586	0.6708	263.51
NSGA-III	0.0796	**0.0029**	**0.0493**	**0.5613**	**257.21**

Table 2. Comparison of the results obtained with the algorithm presented in this work (MO-VND) and NSGA-III over the objective set proposed in ECA.

Method	Avg. HV	Avg. Cov	Avg. IGD+	Avg. G. Spread	Avg. CPUt (s)
MO-VND	0.0672	0.3132	0.3019	0.8657	**155.08**
NSGA-III	**0.0864**	**0.0025**	**0.0894**	**0.7495**	269.71

MO-VND achieves better values in terms of Hypervolume, while NSGA-III has better values in terms of Coverage, IGD+, and Spread on average. Regarding the CPU time used in the process, it can be seen that NSGA-III is a bit faster than MO-VND, although the difference (about 6 s) is small.

In Table 2, we present the results obtained by MO-VND and NSGA-III over the set of objectives described in ECA. Again, we report, for each method, the average Hypervolume (Avg. HV), the average Coverage (Avg. Cov.), the average IGD+ (Avg. IGD+), and the average G. Spread (Avg. Spread) of the generated solutions, in addition to the average CPU time consumed by each method to generate (CPUt (s)). In this case, NSGA-III obtains better results in terms of quality. Interestingly, the CPU time consumed by MO-VND is significantly lower (42.51 %) than the time consumed by NSGA-III.

5   Conclusions

In this work, we have presented an algorithm based on the VNS methodology to tackle the SMCP problem in a multi-objective context. In particular, we proposed a Multi-Objective Variable Neighborhood Descent method with three different neighborhoods. The proposed approach was compared with the well-known NSGA-III for two distinct sets of objectives: MCA and ECA. The results showcase that MO-VND appears to be competitive with NSGA-III in terms of quality of the solutions and computing time. Multi-objective optimization problems in general and the SMCP problem in particular have traditionally been tackled with evolutionary approaches. However, these results indicate that trajectory-based approaches are viable and might be competitive in this context.

The experiments performed to evaluate the algorithmic proposal were carried over a small dataset. Although the instances used were real software projects, a larger dataset is needed to obtain more robust conclusions about the performance of the different methods. This issue will be tackled in future research. Additionally, the results obtained with the ECA objective set suggest that the algorithmic proposal needs new mechanisms to explore the search space. In this regard, another possible extension of this work would be to explore new neighborhoods or to introduce a shake procedure.

Acknowledgements. This research has been partially supported by grants: PID2021-125709OA-C22 and PID2021-126605NB-I00, funded by MCIN/AEI/10.13039/501100011033 and by 'ERDF A way of making Europe'; grant P2018/TCS-4566, funded by Comunidad de Madrid and European Regional Development Fund; grant CIAICO/2021/224 funded by Generalitat Valenciana; and grant M2988 funded by Proyectos Impulso de la Universidad Rey Juan Carlos 2022.

References

1. Arasteh, B., Fatolahzadeh, A., Kiani, F.: Savalan: multi objective and homogeneous method for software modules clustering. J. Softw.: Evol. Process **34**(1), e2408 (2022)
2. Barros, M.D.O.: An analysis of the effects of composite objectives in multiobjective software module clustering. In: Proceedings of the 14th annual conference on Genetic and evolutionary computation, pp. 1205–1212 (2012)
3. Brandes, U., et al.: On modularity clustering. IEEE Trans. Knowl. Data Eng. **20**(2), 172–188 (2007)
4. Chen, C., Alfayez, R., Srisopha, K., Boehm, B., Shi, L.: Why is it important to measure maintainability and what are the best ways to do it? In: 2017 IEEE/ACM 39th International Conference on Software Engineering Companion (ICSE-C), pp. 377–378. IEEE (2017)
5. Chhabra, J.K., et al.: TA-ABC: two-archive artificial bee colony for multi-objective software module clustering problem. J. Intell. Syst. **27**(4), 619–641 (2018)
6. Colanzi, T.E., Assunção, W.K., Vergilio, S.R., Farah, P.R., Guizzo, G.: The symposium on search-based software engineering: past, present and future. Inf. Softw. Technol. **127**, 106372 (2020)
7. Duarte, A., Pantrigo, J.J., Pardo, E.G., Mladenovic, N.: Multi-objective variable neighborhood search: an application to combinatorial optimization problems. J. Global Optim. **63**(3), 515–536 (2015)
8. Durillo, J.J., Nebro, A.J.: jMetal: a java framework for multi-objective optimization. Adv. Eng. Softw. **42**(10), 760–771 (2011)
9. Harman, M., Mansouri, S.A., Zhang, Y.: Search-based software engineering: trends, techniques and applications. ACM Comput. Surv. (CSUR) **45**(1), 1–61 (2012)
10. Huang, J., Liu, J., Yao, X.: A multi-agent evolutionary algorithm for software module clustering problems. Soft. Comput. **21**(12), 3415–3428 (2017)
11. International Organization for Standardization: ISO/IEC/IEEE 24765:2017 Systems and software engineering - Vocabulary (2017)
12. Kumari, A.C., Srinivas, K.: Hyper-heuristic approach for multi-objective software module clustering. J. Syst. Softw. **117**, 384–401 (2016)
13. López-Ibáñez, M., Dubois-Lacoste, J., Cáceres, L.P., Birattari, M., Stützle, T.: The Irace package: iterated racing for automatic algorithm configuration. Oper. Res. Perspect. **3**, 43–58 (2016)
14. Mancoridis, S., Mitchell, B.S., Rorres, C., Chen, Y.F., Gansner, E.R.: Using automatic clustering to produce high-level system organizations of source code. In: 6th International Workshop on Program Comprehension (IWPC 1998), pp. 45–52. IEEE (1998)

15. Martín, R., Cavero, S.: MORK: metaheuristic optimization framewoRK. https://doi.org/10.5281/zenodo.6241738
16. Mitchell, B.S., Mancoridis, S.: A Heuristic Search Approach to Solving the Software Clustering Problem. Drexel University Philadelphia, PA, USA (2002)
17. Mitchell, B.S., Mancoridis, S.: Using heuristic search techniques to extract design abstractions from source code. In: Proceedings of the 4th Annual Conference on Genetic and Evolutionary Computation, pp. 1375–1382 (2002)
18. Mkaouer, W., et al.: Many-objective software remodularization using NSGA-III. ACM Trans. Softw. Eng. Methodol. (TOSEM) **24**(3), 1–45 (2015)
19. Mladenović, N., Hansen, P.: Variable neighborhood search. Comput. Oper. Res. **24**(11), 1097–1100 (1997)
20. Monçores, M.C., Alvim, A.C.F., Barros, M.O.: Large neighborhood search applied to the software module clustering problem. Comput. Oper. Res. **91**, 92–111 (2018)
21. Mu, L., Sugumaran, V., Wang, F.: A hybrid genetic algorithm for software architecture re-modularization. Inf. Syst. Front. **22**(5), 1133–1161 (2020)
22. de Oliveira Barros, M., de Almeida Farzat, F., Travassos, G.H.: Learning from optimization: a case study with apache ant. Inf. Softw. Technol. **57**, 684–704 (2015)
23. Pinto, A.F., de Faria Alvim, A.C., de Oliveira Barros, M.: ILS for the software module clustering problem. XLVI Simpósio Brasileiro de Pesquisa Operacional. Salvador:[sn], pp. 1972–1983 (2014)
24. Pourasghar, B., Izadkhah, H., Isazadeh, A., Lotfi, S.: A graph-based clustering algorithm for software systems modularization. Inf. Softw. Technol. **133**, 106469 (2021)
25. Praditwong, K.: Solving software module clustering problem by evolutionary algorithms. In: 2011 Eighth International Joint Conference on Computer Science and Software Engineering (JCSSE), pp. 154–159. IEEE (2011)
26. Praditwong, K., Harman, M., Yao, X.: Software module clustering as a multi-objective search problem. IEEE Trans. Softw. Eng. **37**(2), 264–282 (2010)
27. Prajapati, A.: Software module clustering using grid-based large-scale many-objective particle swarm optimization. Soft Comput., 1–22 (2022)
28. Prajapati, A., Chhabra, J.K.: A particle swarm optimization-based heuristic for software module clustering problem. Arab. J. Sci. Eng. **43**(12), 7083–7094 (2018)
29. Ramirez, A., Romero, J.R., Ventura, S.: A survey of many-objective optimisation in search-based software engineering. J. Syst. Softw. **149**, 382–395 (2019)
30. Yuste, J., Duarte, A., Pardo, E.G.: An efficient heuristic algorithm for software module clustering optimization. J. Syst. Softw. **190**, 111349 (2022)

An Effective VNS for Delivery Districting

Ahmed Aly[1]([✉]) [iD], Adriana F. Gabor[2,4] [iD], and Nenad Mladenovic[3,4] [iD]

[1] AHOY DMCC: Dubai, Dubai, UAE
ahmed.oss.aly@gmail.com
[2] Applied Mathematics, Khalifa University, Abu Dhabi, UAE
adriana.gabor@ku.ac.ae
[3] Department of Industrial and Systems Engineering, Khalifa University,
Abu Dhabi, UAE
nenad.mladenovic@ku.ac.ae
[4] Research Center on Digital Supply Chain and Operations Management,
Khalifa University, Abu Dhabi, UAE

Abstract. This paper deals with the Delivery Territory Design Problem (DTDP), in which n points have to be allocated to p territories, such that balancing and path connectivity requirements are satisfied, while minimizing the maximum diameter over the created territories. The model is inspired by tactical planning situations faced by delivery companies. We propose two best improvement local search procedures and a Basic Variable Neighborhood Search algorithm following the LIMA paradigm. The results suggest that our algorithm is able to find high-quality solutions within a relatively low time.

Keywords: Territory design · Basic VNS · Less-is-more approach

1 Introduction

The territory design problem *(TDP)*, sometimes referred to as the districting problem or territory alignment problem, is the problem of grouping small geographical units called basic units BUs into larger geographical clusters, named territories, according to relevant planning criteria. Typical applications of the territory design problem include sales territory design [1], political districting [2], school districting [3], and public services districting [4,5]. TDP and its variations have been researched since the 1960s [6], using a variety of models and algorithms. The territory design problem is NP-Hard [7] and thus metaheuristics were used in order to solve this problem. For state-of-the-art models, algorithms, and applications to the territory design problem we refer the reader to [8].

An essential criterion in the TDP is the compactness of districts. One way to achieve this is by minimizing a dispersion measure. A common dispersion measure used in classical problems such as p-median and p-center, is the distance to the centroid of the district. Using a center based measure has some limitations

---

A. Aly—The work of the first author was performed while being a research assistant at Khalifa University.

A. Sleptchenko et al. (Eds.): ICVNS 2022, LNCS 13863, pp. 69–81, 2023.
https://doi.org/10.1007/978-3-031-34500-5_6

for problems where it is not clear how to define possibly "good" centers. In such cases, a dispersion measure based on the diameter of a territory is more appropriate [9].

In this paper, we address a special version of the TDP, called the Delivery Territory Design Problem (DTDP). DTDP aims to design a set of p territories such that the maximum diameter of a territory is minimized, while satisfying several planning requirements such as disjoint territories, and balancing in terms of three attributes: driver workload, commodity demand, and number of customers. Unlike in the TDP, we allow two nodes to be assigned to the same territory as long as there exists a path between them, not necessarily fully contained in the district. The problem is motivated by real-world applications from delivery companies, where a driver can be assigned two serve two BUs as long as there is a street between them.

The rest of this paper is structured as follows; In Sect. 2, we discuss the related works to the territory design problems and their applications. In Sect. 3, we describe the problem and provide the mathematical model of the TDP, while in Sect. 4 we outline the VNS heuristic and its components. We present computational experiments and results in Sect. 5 and conclude our findings in Sect. 6.

2   Related Work

The problem studied in this paper is related to the commercial territory design problem *(CTDP)* which was proposed by Rios-Mercado and Escalante [7]. CTDP seeks to maximize a compactness criterion of p territories subject to planning criteria such as disjoint districts, attribute balancing, and district connectivity. In their paper, compactness is measured by the distance of a node from the center it is assigned to. The authors propose a GRASP algorithm consisting of three phases: construction, adjustment, and local search. The algorithm produces good quality solutions, however that comes at a high computational cost. Rios-Mercado et. al. [9] expanded upon the CTDP with a new model that makes use of a diameter-based dispersion measure instead of its center-based counterpart. They used the GRASP metaheuristic in combination with path relinking to solve instances of 500 nodes and $p = 10$ districts. The algorithm provides good results in terms of the dispersion measure however they are computationally expensive.

In this paper, we propose to solve the DTDP by a Variable Neighborhood Search *(VNS)* algorithm. The core paradigm behind VNS is to systematically change neighborhood structures to prevent plateaus at local optima [10]. Over the years, VNS has been extensively researched and now boasts a wide array of extensions [11]; General VNS, Variable Neighborhood Descent, and Reduced VNS to name a few.

VNS has been successfully used to solve related problems to TDP. Mladenovic et al. [12] proposed a Basic VNS metaheuristic with vertex substitution local search to solve the p-center problem. Their results show that VNS, on average, outperforms Tabu Search, whereas Tabu Search is better for a small p.

Hindi and Fleszar [13] proposed to solve the capacitated p-median problem by a VNS metaheuristic based on the generalized assignment problem. They test their results on five standardized sets of benchmark instances and show that the proposed heuristic finds the best known solutions as well as it improves a previously best-known solution.

Brimberg et. al. [14] showed that Skewed General VNS performs well for the capacitated clustering problem. They showed that evaluating moves prior to accepting inferior solutions is preferable to random shaking procedures. The authors tested the algorithm on the largest set of instances (MDG2000). The Skewed General VNS showed to be the fastest procedure out of the tested meta-heuristics with up to 1.55% improvement over other metaheuristics.

Mladenovic et. al. successfully used the Basic VNS to solve the obnoxious p-median problem in the Less-is-more approach (LIMA) [15]. They proposed a simple facility best improvement local search that lies in between the first and best improvement strategies. They found new best solutions for four instances and ties with 133 instances out of a set of 144 benchmark instances.

Contribution: In this work we propose a VNS based meta-heuristic for solving the DTDP problem, with relaxed connectivity criteria. To the best of our knowledge, this technique has not been previously applied to this problem. Via numerical experiments, we show that the VNS procedure outperforms the algorithm of [9] by 6.35% on average.

3   Problem Description and Mathematical Model

The input to the DTDP is a graph $G = (V, E)$, where the nodes are the set of basic units (BU) and the edges represent the streets between BUs. For each node, we are given a set of attributes A such as number of customers, product demand, and workload. The value of attribute $a \in A$ of BU $i \in V$ will be denoted by w_a^i.

We denote a p-partition of the set V by $X = (X_1, ..., X_p)$ where $X_m \subset V$ is called a territory of V. The size of the territory X_m with respect to attribute $a \in A$ is denoted by $w^a(X_m) = \sum_{i \in X_m} w_i^a$.

We call a partition X *balanced* w.r.t an attribute a, if the size of each territory X_m in X satisfies $\frac{w^a(X_m)}{\mu^a} \in [1 - \tau^a, 1 + \tau^a]$, where μ^a is the average of attribute a over all the nodes. Here, τ^a is a tolerance parameter that is prespecified by the user.

The goal of DTDP is to find a balanced $p-$ partition w.r.t. each attribute, that minimizes the maximum diameter over the territories created. Moreover, we require that any two BUs in a territory are connected by a path in G.

We denote the collection of all the p-partitions of the set V by Π. The TDTP can be formulated as a combinatorial optimization as follows:

$$\min_{X \in \Pi} \max_{m \in M} \max_{i,j \in X_m} \{d_{ij}\} \tag{1}$$

s. t.

$$\frac{w^a(X_m)}{\mu^a} \in [1 - \tau^a, 1 + \tau^a] \quad m \in M, a \in A \tag{2}$$

$$G_m = G(V_m, E) \text{ is connected} \qquad m \in M \tag{3}$$

The formulation in this paper is based on the mathematical model outlined for the CTDP (see Rios-Mercado and Escalante [9]). The objective function (1) minimizes the maximum diameter in a $p-$partition X. Constraints (2) requires that the $p-$partition should be balanced in each attribute $a \in A$. Constraints (3) stipulate that each node of a district must be connected by a path in graph G.

4 Variable Neighborhood Search Procedure

We propose to solve the DTDP by a Basic VNS (BVNS) procedure, in the spirit of the LIMA paradigm [15]. One of the reasons the BVNS is used is that it does not require high computational resources and provides high-quality solutions. Using this variant, we are able to diversify the solution through random neighborhood structures and intensify it through the deterministic neighborhood structures. By doing so, we are able to avoid plateauing at local optima.

The next subsections will outline the construction of the initial solution, the Basic VNS procedure, the local search variants, and the shaking procedure.

4.1 Initial Solution

To generate the initial solution, we use the construction phase of the GRASP algorithm outlined in [9] once. The construction phase starts with a set of p randomly chosen seeds in V. The algorithm then greedily assigns nodes $i \in V$ to the p seeds while attempting to maintain the balancing criteria. If it is not possible to maintain the balancing constraints, the unassigned nodes are allocated to the closest seed. We denote the solution obtained by X_{in}.

4.2 Basic Variable Neighborhood Search

Consider a p-partition $X = (X_1, ..., X_p)$. We define a *neighborhood* $S_k(X)$ as the set of solutions obtained by reallocating k nodes from a territory X_{m_1} to a territory X_{m_2}.

To evaluate the quality of a solution $X = \{X_1, \dots, X_p\}$, the VNS procedure uses the function $\Psi(X)$ introduced in [9] and defined as a linear combination between a function related to the maximum diameter and a measure of the infeasibility of X. More precisely

$$\Psi(X) = \lambda F(X) + (1 - \lambda)G(X),$$

where

$$F(X) = \left(\frac{1}{d_{max}}\right) \max_{m \in M} \max_{i,j \in X_m} \{d_{ij}\},$$

$$d_{max} = \max_{i,j \in V} \{d_{ij}\},$$

and

$$G(X) = \sum_{m=1}^{p} \sum_{a \in A} g^a(X_m),$$

where,

$$g^a(X_m) = \frac{1}{\mu^a} max\{w^a(X_m) - (1 + \tau^a)\mu^a, (1 - \tau^a)\mu^a - w^a(X_m), 0\}.$$

Furthermore, λ is a user specified parameter that controls whether the objective function or the infeasibility is more favored in the cost function calculation.

A general outline of the Basic Variable Neighborhood Search (BVNS) procedure is given in Algorithm 1. The BVNS uses a shake procedure and two local search procedures, called LS-NBI and LS-DBI, that will be described in Sect. 4.3.

Algorithm 1. BVNS($X_{in}, k_{max}, \beta_{max}$)

1: $\beta \leftarrow 1$
2: $X \leftarrow X_{in}$
3: **while** $\beta \leq \beta_{max}$ **do**
4:      $k \leftarrow 1$
5:      **while** $k \leq k_{max}$ **do**
6:          $X' \leftarrow$ **Shake**(X, k)
7:          $X'' \leftarrow$ **LS − NBI**(X')
8:          **if** $\Psi(X'') < \Psi(X)$ **then**
9:              $k \leftarrow 1$
10:             $X \leftarrow X''$
11:         **else**
12:             $k \leftarrow k + 1$
13:         **end if**
14:     **end while**
15: **end while**
16: $X \leftarrow$ **LS − DBI**(X)

The BVNS takes in as input the initial solution X_{in}, the maximum number of neighborhoods used in the shaking procedure k_{max}, and the maximum number of repetitions β_{max}. While the number of repetitions is not reached, the algorithm executes a shake procedure followed by the the local search procedure LS-NBI (lines 6–7). If the value of Ψ is improved, the algorithm performs a sequential neighborhood change step (lines 8–13). Lines (14–15) lead to repeating the BVNS procedure if k_{max} is reached. Finally line (16) applies the second local search variant, LS-DBI.

4.3   Local Search and Shaking Procedures

The BVNS uses two best improvement local search procedures: Node Best Improvement Local Search (LS-NBI) and District Best Improvement Local Search (LS-DBI).

In both procedures, a $move(m,i)$ is defined as re-allocating a node $i \in V \setminus \{X_m\}$ to territory X_m.

Node Best Improvement Local Search (LS-NBI). A pseudo-code for Node Best Improvement (LS-NBI) is given in Algorithm 2. The LS-NBI procedure iterates over all the territories of a solution X. For each territory m, $move(m,i)$, $i \in V \setminus \{X_m\}$ that leads to the best improvement in Ψ is performed (lines 5–11). LS-NBI terminates either when the maximum number of moves is reached (line 3) or no improved solution is found (lines 12–18).

Algorithm 2. LS-NBI(X)

1: $nmoves \leftarrow 0$
2: $optima \leftarrow False$
3: **while** $nmoves < max\_moves$ and $optima = False$ **do**
4:     $improvement \leftarrow False$
5:     **for all** $m \in \{1, \ldots, p\}$ **do**
6:         Find $move(m,i)$ that leads to best improvement of Ψ
7:         **if** Ψ is improved **then**
8:             Perform $move(m,i)$
9:             $improvement \leftarrow True$
10:         **end if**
11:     **end for**
12:     **if** $improvement \leftarrow True$ **then**
13:         $nmoves = nmoves + 1$
14:         $optima = False$
15:     **else**
16:         $optima = True$
17:     **end if**
18: **end while**

District Best Improvement Local Search (LS-DBI). Algorithm District Best Improvement (LS-DBI) is similar to LS-NBI with the main difference being that it iterates over all nodes $i \in V$ instead of territories. For each node i, it performs the best $move(m,i)$, where district m is such that $i \notin X_m$. The procedure terminates either when the maximum number of moves is reached or when no improved solution is found. We refer to Algorithm 3 for the detailed pseudo-code.

Algorithm 3. LS-DBI(X)

1: $nmoves \leftarrow 0$
2: $optima \leftarrow False$
3: **while** $nmoves < max\_moves$ and $optima = False$ **do**
4:    $improvement \leftarrow False$
5:    **for all** $i \in V$ **do**
6:       Find $move(m, i)$ that leads to best improvement of Ψ
7:       **if** Ψ is improved **then**
8:          Perform $move(m, i)$
9:          $improvement \leftarrow True$
10:      **end if**
11:   **end for**
12:   **if** $improvement \leftarrow True$ **then**
13:      $nmoves = nmoves + 1$
14:      $optima = False$
15:   **else**
16:      $optima = True$
17:   **end if**
18: **end while**

Shaking Procedure. The shake procedures serves to diversify the search space. $Shake(X, k)$ chooses two random territories X_1, X_2 of the current solution X and moves k random nodes from X_1 to X_2.

Algorithm 4. Shake(X, k)

1: Choose two random districts $X_1, X_2 \in X$
2: Choose a set K of random nodes in X_1, $|K| = k$
3: Remove K from X_1 and re-allocate the nodes in K to X_2

---

5 Computational Experiments

In this section, we present the results of the numerical experiments we have performed in order to test the proposed algorithms. The algorithms discussed in this section were coded in Python 3.9, and all of the experiments were run on Intel® Xeon X5650 2.67GHz with 72GB RAM.

The computational experiments were performed on randomly generated planar graphs consisting of 500 nodes. We started with a grid graph of 30×30 nodes, divided into 7 regions as in Fig. 1. We generated three types of graphs; Graph Type Center, Graph Type Diagonal, and Graph Type Corners (G-C, G-D, G-CN) by randomly removing $900 - n$ nodes where $n = 500$ from certain regions while maintaining the connectivity of the nodes in the graph.

We remove $\lfloor \frac{3}{4}(900 - n) \rfloor$ nodes from the regions R3, R4, and R5 for G-C; R1, R4, and R7 for G-D; R1, R2, R6, and R7 for G-CN. Finally, we remove

$\lfloor \frac{1}{4}(900 - n) \rfloor$ from the remaining regions for each graph type. When a node is removed, its adjacent edges are also removed.

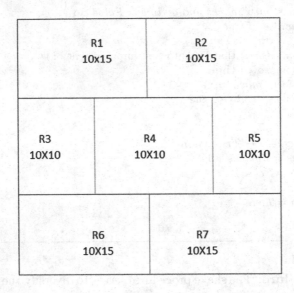

Fig. 1. Graphical representation of the regions of the graph.

Each node in a graph G has three attributes: demand, workload, and number of customers. The attributes were generated from a uniform distribution in the ranges of $[15, 400]$, $[15, 100]$, and $[4, 20]$ respectively [7]. Furthermore, to simulate real-world scenarios, we generated a distance attribute for edges from a uniform distribution in the range of $[5, 20]$. The value of the parameters was chosen as follows. For the cost function $\Psi(X)$, parameter λ was set to $\lambda = 0.7$. Furthermore, $p = 10$, $\beta_{max} = 5$, $m_{max} = 4$, and $max\_moves = 100$.

5.1   Impact of Local Search on Initial Solution

In Table 1, the columns LS-NBI and LS-DBI refer to the algorithms in which the initial solutions are improved by applying the respective local search once. Furthermore, the column labeled with "GRASP-LS" refers to the local search procedure described in [9].

The average improvement of LS-NBI, LS-DBI, and GRASP in terms of the objective function value over the initial solutions was 4.00%, 4.32%, and 1.59% respectively. Table 1 shows the average percentage improvement of the local search variants on each graph type. We note that LS-DBI outperformed the rest of the variants over graph types G-CN and G-D at 5.26% and 3.81% respectively. While LS-NBI, outperformed the rest of the variants for graph type G-C at 4.72%.

Table 1. Local search procedures percentage improvement over the initial solution.

Graph		LS-NBI	LS-DBI	GRASP-LS
Graph-Type	Measure			
G-C	*Max*	14.29%	9.13%	5.34%
	Average	4.72%	3.88%	1.18%
	Min	0%	0%	0%
G-D	*Max*	9.53%	12.61%	7.16%
	Average	2.97%	3.81%	2.43%
	Min	0%	0%	0%
G-CN	*Max*	14.56%	14.56%	4.98%
	Average	4.31%	5.26%	1.15%
	Min	0%	0%	0%

We observe that the local search variants LS-NBI and LS-DBI have a different impact based on the graph type that the variant was performed on. Furthermore, we notice that both local search procedures had a high variance in graph type G-CN. This suggests that this particular graph type is difficult to improve upon.

We note that the local search variant LS-DBI outperformed LS-NBI and GRASP-LS for graph types G-D and G-CN at 3.81% and 5.26% respectively. On the other hand, LS-NBI outperformed all other variants in graph type G-C at 4.72%. Due to the different performance of both LS-NBI and LS-DBI based on the graph type, we used both local search variants in $BVNS(X, k_{max}, \beta_{max})$.

5.2 Computational Experiments on BVNS

We compared the results of the proposed BVNS algorithm on all graph types with the results of the static Path-Relinking (PR) algorithm presented in [9].

Figure 2 shows an example solution and a comparison between BVNS and PR. Figure 2(a) shows the BVNS solution of that particular instance while Fig. 2(b) shows the PR solution where each have ten distinct districts. We can see that due to the shaking procedure of the BVNS algorithm, the solution was able to escape local optima regardless of the graph structure and provide significant improvements.

Table 2. Objective Function Percentage Improvement of BVNS over PR.

Graph	Min	Max	Average
G-C	0.95%	18.62%	6.42%
G-D	0%	13.20%	5.21%
G-CN	0%	20.86%	7.42%
Average			**6.35%**

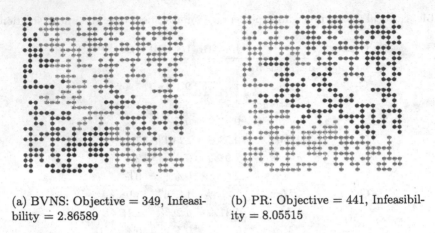

(a) BVNS: Objective = 349, Infeasi-   (b) PR: Objective = 441, Infeasibil-
bility = 2.86589   ity = 8.05515

Fig. 2. Example comparison of BVNS and PR.

Table 2 summarizes the results of the objective function improvement of BVNS over PR. We show the minimum, maximum, and average improvement of BVNS over PR for each graph type. On average, BVNS outperforms PR by 6.35% over all graph types with a maximum improvement of 20.86% for G-CN. On average, BVNS outperformed PR by 6.42% for graph type G-C, 5.21% for graph type G-D, and by 7.42% for graph type G-CN. We note that BVNS outperformed PR for 27 out of the 30 graphs. PR outperformed BVNS for 3 out of the 30 graphs in terms of objective function value. We note that 2 out of the 3 graphs where PR outperformed BVNS occurred in graph type G-D.

Figure 3 shows the box plot of the percentage improvement of the relative infeasibility of BVNS over PR. We can see that the relative infeasibility of BVNS has shown consistent improvements over PR with the certain outliers in each graph type. This indicates that using the BVNS procedure with LS-NBI and LS-DBI led to improvements of the infeasibility of the solution compared to PR, on average by 21.7%. The relative infeasibility in graph type G-CN has been the most unstable where 3 out of the 6 instances show PR outperforming BVNS in terms of relative infeasibility.

Furthermore, Fig. 3 along with the results presented for the average objective function improvement, suggest that the graph type G-D is difficult to improve for both algorithms.

5.3   Running Times

The average running time of BVNS was 394.49 s over all graph types. Among the local search procedures, LS-DBI was less time consuming, with an average of 13.17 s, followed by LS-NBI with an average running time of 17.24 s. Thus, the time difference between the two local search procedures is negligible and provides great benefit in tackling different types of graphs with considerable improvements.

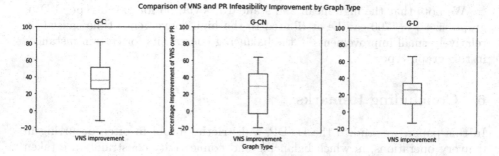

Fig. 3. Relative Infeasibility of BVNS and PR.

To examine the scalability of the algorithm, we tested BVNS on graphs with nodes $V = \{500, 600, 700\}$ with 10 graphs for each graph type. Table 3 shows the CPU time (in mins.) and the standard deviation of every graph type at a different number of nodes.

Table 3. CPU Time (in sec.) and the Standard Deviation of Each Graph Type.

Graph		500 Nodes	600 Nodes	700 Nodes
Graph Size	CPU Time			
G-C	*Mean*	405.12	821.50	941.52
	SD	103.49	444.32	243.47
G-CN	*Mean*	369.77	843.76	869.22
	SD	75.55	242.31	227.85
G-D	*Mean*	408.58	627.08	1117.42
	SD	125.19	169.98	363.40

As expected, the running time increases as the problem size grows larger at $V = 100$ increments. We note that for graph sizes $V = 500$ and $V = 700$, graph type G-CN had the lowest mean and standard deviation in their respective graph sizes. On the other hand, for graph size $V = 600$, graph type G-D had the lowest mean and standard deviation.

Furthermore, we can see that in the first increment between $V = 500$ and $V = 600$, the mean CPU time increased at a higher rate that between $V = 600$ and $V = 700$ for graph types G-C and G-CN. This suggests that graph types G-C and G-CN scale well as the problem increases in size. In addition, we note that for graph type G-D, the mean and standard deviation had their lowest rate of increase between graph sizes $V = 500$ and $V = 600$. This suggests that for this particular graph type, scaling the problem size up to $V = 600$ would not present significant increases in run time and that the algorithm would provide a solution in an adequate timeframe.

We note that the high mean and standard deviation for graph type G-D in graph size $V = 700$ can be attributed to the high convergence time caused by relatively small improvements in the balancing constraints for certain instances in this graph type.

6   Concluding Remarks

In this paper, we studied the DTDP, a districting problem often occurring in delivery operations, in which balancing and connectivity constraints are taken into account while minimizing the maximum diameter. We proposed two local search procedures that improve the objective function value and lower the relative infeasibility of a given solution. We used a Basic VNS following the LIMA paradigm under which we used both the local search variants and a simple shake procedure. We conducted computational experiments on graphs of $V = 500$ nodes and $p = 10$ districts with a competitive running time and average improvement at 6.35% over all graph types and a maximum improvement of 20.86%. Furthermore, we conducted computational experiments on graphs of $V = 600$ and $V = 700$ to showcase the scalability of the algorithm.

There are several areas of future research that arise from this problem. One promising area of research, given the results of the Basic VNS, is exploring other variants of VNS such as the General VNS.

Furthermore, applying different local search neighborhoods in conjunction with different neighborhood change steps could allow for further diversification and intensification of the solution space.

Acknowledgements. This research is supported by Khalifa University under Grant No. FSU-2020-19 and Award No. RC2 DSO.

References

1. Zoltners, A.A., Sinha, P.: Sales territory design: thirty years of modeling and implementation. Mark. Sci. **24**(3), 313–331 (2005)
2. Ricca, F., Scozzari, A., Simeone, B.: Political districting: from classical models to recent approaches. Ann. Oper. Res. **204**(1), 271–299 (2013)
3. Caro, F., Shirabe, T., Guignard, M., Weintraub, A.: School redistricting: embedding GIS tools with integer programming. J. Oper. Res. Soc. **55**(8), 836–849 (2004)
4. Enayati, S., Mayorga, M.E., Rajagopalan, H.K., Saydam, C.: Real-time ambulance redeployment approach to improve service coverage with fair and restricted workload for EMS providers. Omega (Westport) **79**, 67–80 (2018)
5. Sudtachat, K., Mayogra, M.E., Mclay, L.A.: A nested-compliance table policy for emergency medical service systems under relocation. Omega (Westport) **58**, 154–169 (2016)
6. Sandoval, G.M., Diaz, J.A., Rios-Mercado, R.: An improved exact algorithm for a territory design problem with p-center-based dispersion minimization. Expert Syst. Appl. **146**, 113150 (2020)

7. Rios-Mercado, R.: Fernandez, E: a reactive GRASP for a commercial territory design problem with multiple balancing requirements. Comput. Oper. Res. **36**(3), 755–776 (2009)
8. Kalcsics, J., Nickel, S., Schröder, M.: Towards a unified territorial design approach: applications, algorithms, and GIS integration. TOP **13**(1), 1–56 (2005)
9. Rios-Mercado, R., Escalante, H.: GRASP with path relinking for commercial districting. Expert Syst. Appl. **44**, 102–113 (2015)
10. Mladenović, N., Hansen, P.: Variable neighborhood search. Comput. Oper. Res. **24**(11), 1097–1100 (1997)
11. Hansen, P., Mladenovic, N., Todosijevic, R., Hanafi, S.: Variable neighborhood search: basics and variants. EURO J. Comput. Optim. **5**, 423–454 (2016)
12. Mladenovic, N., Labbe, M., Hansen, P.: Solving the p-center problem with Tabu search and variable neighborhood search. Networks **42**(1), 48–64 (2003)
13. Hindi, K.S., Fleszar, K.: An effective VNS for the capacitated p-median problem. Eur. J. Oper. Res. **191**, 612–622 (2008)
14. Brimberg, J., Mladenovic, N., Todosijevic, R., Uroševic, D.: Solving the capacitated clustering problem with variable neighborhood search. Ann. Oper. Res. **272**, 289–321 (2019)
15. Mladenovic, N., Alkandari, A., Pei, J., Todosijevic, R., Pardalos, M.P.: Less is more approach: basic variable neighborhood search for the obnoxious p-median problem. Intl. Trasn. in Op. Res. **27**(1), 1–14 (2019)

BVNS for the Minimum Sitting Arrangement Problem in a Cycle

Marcos Robles[ID], Sergio Cavero[✉][ID], and Eduardo G. Pardo[ID]

Universidad Rey Juan Carlos, Madrid, Spain
{marcos.robles,sergio.cavero,eduardo.pardo}@urjc.es

Abstract. Signed graphs are a particular type of graph, in which the vertices are connected through edges labeled with a positive or negative weight. Given a signed graph, the Minimum Sitting Arrangement (MinSA) problem aims to minimize the total number of errors produced when the graph (named the input graph) is embedded into another graph (named the host graph). An error in this context appears every time that a vertex with two adjacent (one positive and one negative) is embedded in such a way that the adjacent vertex connected with a negative edge is closer than the adjacent vertex connected with a positive edge. The MinSA can be used to model a variety of real-world problems, such as links in online social networks, the location of facilities, or the relationships between a group of people. Previous studies of this problem have been mainly focused on host graphs, whose structure is a path. However, in this research, we compare two variants of the MinSA that differ in the graph used as the host graph (i.e., a path or a cycle). Particularly, we adapted a previous state-of-the-art algorithm for the problem based on Basic Variable Neighborhood Search. The solutions obtained are compared with the solutions provided by a novel Branch & Bound algorithm for small instances. We also analyze the differences found by using non-parametric statistical tests.

Keywords: Minimum Sitting Arrangement · Graph embedding · Basic Variable Neighborhood Search · Cycle host graph

1 Introduction

The Minimum Sitting Arrangement (MinSA) problem belongs to a family of combinatorial optimization problems denoted as Graph Layout Problems (GLP). The aim of GLPs is to find a layout of an input graph that optimizes a certain

This research has been partially supported by grants: PID2021-125709OA-C22, PID2021-126605NB-I00 and FPU19/0409, funded by MCIN/AEI/10.13039/501100011033 and by "ERDF A way of making Europe"; grant P2018/TCS-4566, funded by Comunidad de Madrid and European Regional Development Fund; grant CIAICO/2021/224 funded by Generalitat Valenciana; and grant M2988 funded by Proyectos Impulso de la Universidad Rey Juan Carlos 2022. We would also like to thank authors of [21] for sharing their code with us.

objective function. A layout consists of embedding a graph, denoted as input graph, in another graph, denoted as host graph, by mapping the vertices of the input graph to the vertices of the host graph. GLPs have acquired significant interest in the scientific community due to their multiple applications in the real world such as circuit design, telecommunication network migration, facility location, language syntax analysis or graph drawing, among others [4,5,12,26].

Within the GLP family, the most studied problems are those in which the embedding is performed on a path host graph [12,22]. In these variants, practitioners have focused on several well-known objective functions such as: the bandwidth [6,8], the cutwidth [9,14], the minimum linear arrangement [24], or the vertex separation [25]. Furthermore, some of those objective functions have also been studied for other host graphs with regular structure such as: cycle graphs [7,9], tree graphs [13], or grid graphs [6], among others [12].

The MinSA is a GLP originally introduced and studied using a path host graph, but recently adapted to cycle host graphs [3]. Given an input graph with positive and negative edges (that is, labeled with +1/−1), the objective of the MinSA is to find a linear/circular layout where the vertices connected with positive edges are placed closer than the vertices connected with negative edges. In particular, the objective function of MinSA looks for the minimization of the number of negative connections placed closer than a positive one. However, the study of the same objective function but for different regular-structured host graphs, might require very different solution approaches, resulting in different optimization problems.

The MinSA can be related to a wide range of real-world applications, for example, the assignment of frequencies to radio channels, the modeling of relationships or links in social networks, or the location of facilities on a map [1–3,11]. These practical applications are some of the motivating factors for the problem under investigation.

In this paper, we study the differences of the best solutions obtained for the MinSA when the host graph is a path in comparison with a cycle. To this end, we implemented the best previous state-of-the-art heuristic algorithm for the MinSA, based on Basic Variable Neighborhood Search (BVNS), which was proposed for the path, reporting the best solutions found, and we evaluated them when the embedding is performed in a cycle. Then, we modified the algorithm to optimize the MinSA specifically for the cycle, identifying if the performance of the previous algorithm vary. Finally, we compare the best results obtained by the BVNS procedure for the MinSA considering a path and a cycle host graph.

The rest of the paper is organized as follows. In Sect. 2, we review the state of the art of the MinSA. Then, in Sect. 3, we formally define the problem. In Sect. 4 we present the best algorithmic approach for the MinSA. Next, in Sect. 5, we describe the experiments carried out and compare and analyze the results obtained. Finally, in Sect. 6, we provide general conclusions and relevant directions for future research.

2   State of the Art

The MinSA problem was initially introduced in [18], where it was defined for a path host graph. In that work, it was proved that the problem can be solved in polynomial time for complete input graphs. Lately, the problem was proved to be NP-Complete if the input graph is not a complete graph [11]. Therefore, the MinSA problem has been studied from both theoretical [11] and heuristic perspectives [21,23]. From a heuristic perspective, the method proposed in [21] can be considered the current state-of-the-art algorithm for the problem, when defined for a path host graph. Specifically, in that paper, the authors proposed a greedy constructive procedure and a Basic Variable Neighborhood Search algorithm.

Recently, the MinSA problem has been extended by considering the embedding of the candidate graph in other structures. In [3], the authors proposed the embedding of signed input graphs in a cycle. Moreover, in [3], it was proved that given a complete signed graph, there exists an embedding with an associated objective function value equal to 0, if its positive subgraph is a proper circular-arc graph [27]. The MinSA problem has also been theoretically studied by considering tree host graphs. In [2], the authors proved that a complete input graph can be embedded in a tree with 0 errors, if the input graph is strongly chordal [15].

In addition to the previous theoretical results, as far as we are concerned, no general algorithms (either exact or heuristic) have been proposed for the variants of the problem where the host graph is either a cycle or a tree.

3   Problem Statement

In general, a graph layout problem can be formally defined as the embedding of an input graph G in a host graph H, such that an objective function is optimized. The input graph is defined as $G = (V_G, E_G)$, where V_G and E_G represent the sets of vertices and edges of the input graph, respectively, and the host graph is defined as $H = (V_H, E_H)$, where V_H and E_H represent the sets of vertices and edges of the host graph, respectively. The embedding, also known as projection, labeling, or mapping, consists of defining two mathematical functions. The first, usually denoted as φ, is a bijective function that assigns each vertex of the input graph to a vertex of the host graph such that $\varphi : V_G \to V_H$. Therefore, a vertex $u \in V_G$ is assigned to a vertex $v \in V_H$ if $\varphi(u) = v$. The second function, commonly denoted ψ, is an injective function that assigns each edge $(u, v) \in E_G$ to a path in H with endings in $\varphi(u)$ and $\varphi(v)$.

When the input graph $G = (V_G, E_G)$ is a signed graph, E_G has the particularity of being divided into two subsets: $E_G = \{E^+, E^-\}$, with $E^+ \cup E^- = E_G$ and $E^+ \cap E^- = \emptyset$. The subset E^+ contains the edges signed as *positive*. Similarly, the subset E^- contains the edges labeled as *negative*. The sign of an edge can also be understood as a positive or negative weight $(+1/-1)$. For example, in Fig. 1a we show an input signed graph G with $|V_G| = 5$ and $|E_G| = 5$, where the vertices are alphabetically labeled, being $V_G = \{A, B, C, D, E\}$, $E_G = \{(A, B), (A, E), (B, E), (C, E), (D, E)\}$, with $E^+ = \{(B, E), (C, E)\}$ and $E^- = \{(A, B), (A, E), (D, E)\}$.

Among the different regular-structured host graphs for which the MinSA has been studied, in this paper we focus our attention on the comparison between path graphs and cycle graphs. A path graph, denoted as P_n, is a connected graph with n vertices and $n - 1$ edges such that $n = |V_G| = |V_H|$. On the other hand, a cycle graph C_n is a connected graph with n vertices and n edges such that $n = |V_G| = |V_H|$. For example, given the input graph G depicted in Fig. 1a we illustrate a path host graph P_5 in Fig. 1b able to host the vertices of G. Similarly, in Fig. 1c, we illustrate a cycle host graph C_5, suitable to host the vertices of G. In the context of MinSA, the number of vertices in both host graphs must be 5, since $|V_G| = |V_{P_5}| = 5$ and $|V_G| = |V_{C_5}| = 5$. Furthermore, in both examples, the vertices of the host graphs have been labeled with numbers from 1 to 5.

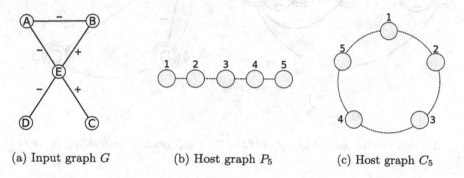

(a) Input graph G (b) Host graph P_5 (c) Host graph C_5

Fig. 1. (a) Example of an input signed graph G with 5 vertices and 5 edges. (b) The path host graph P_5 for graph G. (c) The cycle host graph C_5 for graph G.

Finally, the embedding is performed through the definition of the functions φ and ψ. In particular, φ, as mentioned above, relates each vertex of G to a vertex of H. Then, ψ, in the particular case of the MinSA, assigns the shortest path in H to an edge (u, v) of G. More formally:

$$\psi((u, v)) = \underset{p(\varphi(u), \varphi(v)) \in P_H}{\arg \min} \{|p(\varphi(u), \varphi(v))|\} \, \forall \, (u, v) \in E_G, \qquad (1)$$

where P_H represents the set of all possible paths in H. Note that in the case of a path host graph, there is only one possible path that can be assigned to every input edge. However, in the case of the cycle host graph, there are two possible paths (avoiding loops) between each pair of vertices. Additionally, since ψ can be derived from φ, in order to simplify the notation, in the rest of the document, we only use φ_P and φ_C to denote an embedding (i.e., a solution of the problem) in a path or a cycle host graph, respectively.

Let us illustrate the concept of embedding with examples. In Fig. 2a we show a possible embedding φ'_P of G in the path host graph P_5. In this example, the function φ'_P has been defined as follows: $\varphi'_P(A) = 1$, $\varphi'_P(B) = 2$, $\varphi'_P(C) = 3$, $\varphi'_P(D) = 4$, and $\varphi'_P(E) = 5$. Similarly, in Fig. 2b we illustrate a possible embedding φ'_C of G in a cycle host graph C_5 where φ'_C is defined by the assignments $\varphi'_C(A) = 1$, $\varphi'_C(B) = 2$, $\varphi'_C(C) = 3$, $\varphi'_C(D) = 4$, and $\varphi'_C(E) = 5$.

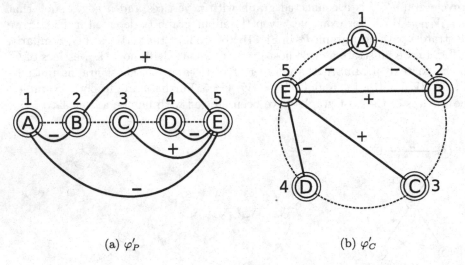

(a) φ'_P (b) φ'_C

Fig. 2. (a) A possible embedding φ'_P of G in P_5. (b) A possible embedding φ'_C of G in C_5.

With the previous definitions at hand, we now introduce the concept of error that is necessary for defining the objective function of the MinSA, either in the path or in the cycle. Given a function φ, three vertices $u, v, w \in V_G$, and two edges, $(u, v) \in E^+$ and $(u, w) \in E^-$, an error in u, denoted as $\mathcal{E}(u, \varphi)$, occurs if $\varphi(w) \in \psi((u, v))$, where $\varphi(w)$ is the host vertex assigned to w.

Then, the number of errors produced in an embedding φ, denoted as $\mathcal{E}(\varphi)$, is calculated as the sum of errors across the entire set of vertices, for each pair of edges with a positive and negative sign, respectively. More formally:

$$\mathcal{E}(\varphi) = \sum_{u \in V_G} \mathcal{E}(u, \varphi). \tag{2}$$

Finally, the objective of MinSA is to find an embedding φ^*, among the set of all possible embeddings Φ, that minimizes the total number of errors. In mathematical terms:

$$\varphi^* = \arg \min_{\varphi \in \Phi} \mathcal{E}(\varphi). \tag{3}$$

To illustrate how to evaluate the objective function of MinSA, considering either a path or a cycle host graph, we again use the embeddings φ'_P and φ'_C, shown in Fig. 2. Furthermore, the number of errors associated with each vertex in G is presented in Table 1. For example, vertex B has one positive edge (with vertex E) and one negative edge (with vertex A). To calculate the number of errors of B on the path host graph, $\mathcal{E}(B, \varphi'_P)$, we determine whether $\varphi'_P(A)$ is contained in the path assigned to the edge (B, E) in the host graph, through $\psi((B,E))$. Since $\psi((B,E)) = \{2, 3, 4, 5\}$, and $\varphi'_P(A) = 1 \notin \{2, 3, 4, 5\}$, no errors are produced. Now, we calculate the number of errors of B in the cycle host graph, $\mathcal{E}(B, \varphi'_C)$. Similarly, we determine whether $\varphi'_C(A)$ is contained in the path assigned to the edge (B, E). In this case, there are two possible paths in the host graph between $\varphi'_C(B)$ and $\varphi'_C(E)$: $\{2, 3, 4, 5\}$ and $\{2, 1, 5\}$. As stated in Eq. 1, the shortest path in the host graph is assigned to each edge, and therefore $\psi((B,E)) = \{2, 1, 5\}$. Since $\varphi'_C(A) = 1 \in \{2, 1, 5\}$, an error is produced.

Finally, given the number of errors produced by each vertex of G in the path host graph (depicted in Fig. 2a), the value of the objective function of the MinSA problem, for the solution φ'_P, is calculated as: $\mathcal{E}(\varphi'_P) = 0 + 0 + 0 + 0 + 2 = 2$. Similarly, given the number of errors produced by each vertex of G in the cycle host graph (depicted in Fig. 2b), the value of the objective function of the MinSA problem, for the solution φ'_C (depicted in Fig. 2b) is $\mathcal{E}(\varphi'_C) = 0 + 1 + 0 + 0 + 2 = 3$. As it can be observed, both solutions do not have the same objective function value despite their similarities.

Table 1. Number of errors of each vertex of G embedded in a path and cycle host graph through functions φ'_P and φ'_C, respectively.

Vertex	#Errors in P_5	#Errors in C_5
A	$\mathcal{E}(A, \varphi'_P) = 0$	$\mathcal{E}(A, \varphi'_C) = 0$
B	$\mathcal{E}(B, \varphi'_P) = 0$	$\mathcal{E}(B, \varphi'_C) = 1$
C	$\mathcal{E}(C, \varphi'_P) = 0$	$\mathcal{E}(C, \varphi'_C) = 0$
D	$\mathcal{E}(D, \varphi'_P) = 0$	$\mathcal{E}(D, \varphi'_C) = 0$
E	$\mathcal{E}(E, \varphi'_P) = 2$	$\mathcal{E}(E, \varphi'_C) = 2$

4 Algorithmic Strategies

In this paper, we study the behavior of a previous heuristic algorithm based on Variable Neighborhood Search designed for the MinSA in the path, when used to generate solutions for the MinSA in the cycle. Furthermore, we compare their results with an exact algorithm, a basic Branch & Bound.

The heuristic algorithm was introduced in [21] and, as far as we know, it is the current state of the art of the problem. It consists of a greedy constructive procedure to generate an initial solution and a Basic Variable Neighborhood Search (BVNS) to improve it.

The constructive procedure proposed in [21] is based on the identification of existent cliques in the input graph, when considering only the positive edges. Then, the vertices belonging to the same clique are placed in the embedding consecutively. The cliques are selected following a descending order of size (in terms of the number of vertices of the clique). Later, the remaining vertices are embedded in any unassigned host vertex at random. This approach is based on the idea that the vertices within the clique are suitable for placement together in the layout because they do not create errors among them.

The solution obtained by the previous constructive method is provided as an initial solution to a BVNS procedure. BVNS is one of the most relevant variants of Variable Neighborhood Search metaheuristic. VNS was proposed by Mladenović and Hansen as a general method to solve hard combinatorial optimization problems [16,17]. The main principle of this methodology is to make systematic changes to the neighborhood structure in order to escape from local optima solutions. The pseudocode of BVNS is presented in Algorithm 1. This procedure, in addition to the initial solution (φ), receives two additional parameters: the maximum computation time (t_{max}) and the maximum number of neighborhoods to be explored (k_{max}). In particular, BVNS is made up of three procedures, Shake (step 5), LocalSearch (step 6) and NeighborhoodChange (step 7). The Shake procedure performs k random movements within the associated neighborhood. In this case, the shake is based on the swap move, a classical move operator that consists of exchanging the assignment of two vertices of the input graph. Then, the LocalSearch procedure finds a local optimum starting from the solution provided by the Shake procedure, within the neighborhood defined by the insert move. The insert move consists of removing an input vertex from its current position in the host graph and inserting it in a different position. As is customary in insertion moves, the displaced elements must be shifted. In this proposal, the vertices are shifted in the direction in which the least number of vertices are affected. Finally, the NeighborhoodChange procedure is responsible for determining whether the new solution under consideration has improved the best solution found in the procedure or not. If so, $k = 1$ and the best solution is updated, otherwise the value of k increases. If a solution is not improved after exploring the neighborhood k_{max} (step 4), then the procedure starts again from $k = 1$, as long as the elapsed time is less than t_{max} (step 2).

To evaluate the previous BVNS algorithm to solve MinSA for a cycle host graph, we have configured two variants of the procedure, denoted as $BVNS_1$ and $BVNS_2$, which differ in the criterion used by each method for considering an improvement move during the search. Note that $BVNS_1$ corresponds to the original implementation of the procedure, defined for the path host graph. Therefore, every time that a move is performed, $BVNS_1$ evaluates the solution as if it were a path, accepting or not the move if an improvement is made considering the evaluation on the path. When the method is not able to further improve the solution with this strategy, it stops and evaluates the final solution obtained over the cycle. On the other hand, $BVNS_2$ modifies $BVNS_1$ by evaluating the resulting solution after a move, considering the solution as a cycle

---

Algorithm 1. State-of-the-art algorithm for the MinSA

1: BVNS($\varphi, t_{max}, k_{max}$)
2: **while** ElapsedTime() $< t_{max}$ **do**
3:     $k \leftarrow 1$
4:     **while** $k < k_{max}$ **do**
5:         $\varphi' \leftarrow$ Shake(φ, k)
6:         $\varphi'' \leftarrow$ LocalSearch(φ')
7:         $\varphi, k \leftarrow$ NeighborhoodChange(φ, φ'', k)
8:     **end while**
9: **end while**
10: **return** φ

(that is, two possible paths can be assigned to each input edge and the shortest one is selected). The comparison between these two methods aims to determine whether an algorithm designed for the MinSA over the path can be directly used for the MinSA over the cycle, or at least it can be easily adapted.

Additionally, we have implemented an exact algorithm, based on Branch & Bound (B&B) [19,20], for comparison purposes. The B&B is an exact strategy designed as an evolution of the Backtracking algorithm [10]. It usually represents the solution space as a tree, where each intermediate node represents a partial solution, and each final node (i.e., the leaves of the tree) represent a complete solution. The algorithm tries the exploration of all nodes of the tree, avoiding those branches which do not lead to promising solutions, using bounds in each node, to avoid wasting computational time. In this case, we introduce a very basic version of B&B where the bounds are based on two strategies: 1) avoiding the exploration of partial solutions with an objective function value equal to or larger than the best overall solution found; and 2) avoiding the exploration of equivalent solutions. Notice, that a solution is equivalent to a previously explored one when the relative order in the host graph between each pair of candidate vertices is equal to a previously explored solution.

5   Experiments

This section is devoted to studying the performance of the algorithms introduced in Sect. 4, when tackling MinSA defined using a cycle host graph. The experiments presented in this section are carried out on a representative subset of 60 instances selected from a previously introduced data set for the problem [21,23]. This set is made of three types of randomly generated instances: Random, Interval and Complete. Random instances, as the name indicates, are composed of random graphs. Interval instances are graphs where the positive edges form a unit interval graph, and negative edges are added on top of the unit interval positive structure. Finally, Complete instances are complete graphs where the positive/negative edges are randomly set. The data set used can be downloaded at https://www.heuristicas.es/publications/minsa-icvns. All experiments have been executed on a virtual CPU AMD EPYC 7282 8-Core and 8 GB of RAM.

Note that we have directly used the original code implemented by the authors of [21], who kindly provided us with the source code of their algorithms for our research. Additionally, all algorithms compared (BVNS$_1$, BVNS$_2$ and the B&B) have been implemented in Java 17 and no other commercial software, such as a solver, has been used.

In Table 2 we report the results obtained by a single execution of the BVNS$_1$, BVNS$_2$ and the B&B over the set of previously introduced instances. The results are grouped by type of instance (Complete, Interval, or Random) and, additionally, at the bottom of the table, we have added a row (labeled as "Total") that provides the average for all instances of the table. In particular, we report the average of the best solutions found (Avg.), the deviation from the best solution found in the experiment (Dev. (%)), the CPU time in seconds (CPU T.(s)), and the number of the best solutions found in the experiment (# Best). Notice that in the case of the B&B, the number of the best solutions found matches the number of optima found by the method.

Analyzing the performance of three algorithms compared, as expected, BVNS$_2$ is the best procedure for the MinSA over the cycle, since it obtains the best average quality of the solutions, the lowest deviation, and the highest number of the best solutions found. Although BVNS$_2$ is just a modification of BVNS$_1$, it is capable of reaching the global optimum for 13 instances (those where the B&B certified the optimum). On the other hand, the computation time of the BVNS$_2$ is an order of magnitude longer than the BVNS$_1$. This can be partially explained by the fact that BVNS$_2$ is able to find better solutions during the search and, therefore, continues the search for longer time. In any case, we can state that the performance of the BVNS$_2$ has considerably improved the original method in terms of quality of the objective function. As expected, the performance of the B&B method decreases when the size of the instances increases. The method was able to find 13 optimum values for small instances; however, for large instances it was only able to reach some feasible solutions of lower quality than BVNS$_1$ and BVNS$_2$ in the time limit of 3600 s.

In the Appendix section, we include the individual results for each instance considered in Table 2.

To complement the previous experiment, we have statistically analyzed the differences between the values of the objective function of the best solutions found by BVNS$_1$ and BVNS$_2$. To this end, we used a non-parametric statistical test, since the samples do not follow a normal distribution. In particular, we employ Wilcoxon's signed rank test [28], a test which objective is to determine whether two samples are likely to derive from the same population (algorithms in this case). The test calculates the difference between pairs and analyzes these differences to determine whether they differ significantly from each other or not. The null hypothesis, in this case, indicates that there is no significant difference between two populations (that is, there is no difference between the solutions found by both algorithms). The obtained p-value less than 0.0001 indicates that we can reject the null hypothesis and confirm the existence of significant differences among the best solutions found.

Table 2. Comparison of the algorithmic strategies described in this paper.

		Avg.	CPU T.(s)	Dev. (%)	# Best
Complete (20)	$BVNS_1$	15978.50	31.68	25.34	2
	$BVNS_2$	14761.55	119.91	0.00	20
	B&B	18893.55	2892.14	33.03	4
Interval (20)	$BVNS_1$	206.40	19.18	106.45	7
	$BVNS_2$	147.75	43.67	7.35	18
	B&B	3077.75	2706.39	2974.34	5
Random (20)	$BVNS_1$	16826.80	29.54	30.26	3
	$BVNS_2$	15658.60	642.99	0.00	20
	B&B	19883.45	2910.45	100.59	4
Total (60)	$BVNS_1$	11003.90	26.80	54.02	12
	$BVNS_2$	10189.30	268.80	2.45	58
	B&B	13951.58	2836.32	1035.99	13

Finally, to observe the relationship between the objective function of the MinSA when the graph is embedded either in the path or in the cycle, we introduce the Fig. 3. In particular, we display three scatter plots, one for each set of instances: complete graphs in Fig. 3a, Interval graphs in Fig. 3b, and Random graphs in Fig. 3c.

Each point in the graphs represents an instance, the X-axis represents the value of the objective function for the MinSA when the host graph is a path, and the Y-axis represents the value of the objective function for the MinSA when the host graph is a cycle. In each chart, we have drawn a dashed straight line that comfortably fits through the data; hence, we can observe the existence of a linear relationship between both objective functions. The slope of the line is positive, so there is a positive correlation between the value of the objective function of both problems. Interestingly, since the slope is greater than 1 in the three charts (1.36 in complete graphs, 1.14 in interval graphs, and 1.32 in random graphs) we can state that the value of the objective function for an instance of a MinSA solution is generally higher in the case of the cycle than in the case of the path. This may be useful for future researchers as a starting point for studying the relationship between both problems and derive lower/upper bounds between them.

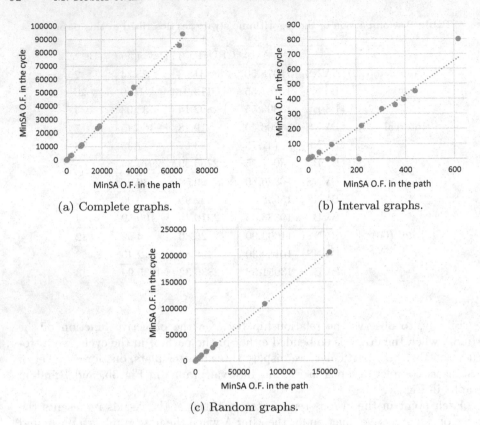

(a) Complete graphs.

(b) Interval graphs.

(c) Random graphs.

Fig. 3. Relation between the quality of the objective function of the MinSA problem when the host graph is a path or a cycle for different sets of instances.

6   Conclusions

In this paper we have studied the Minimum Sitting Arrangement Problem, that consists of embedding an input graph into a host graph while minimizing the sum of the number of errors in the embedding. Particularly, we tackled two variants of the problem depending on the structure of the host graph: a path or a cycle.

The problem has been addressed using exact and heuristic algorithms. The exact proposal consists of a Branch & Bound algorithm that reaches optimal solutions for small instances in a few seconds. However, for large instances, it is only able to reach feasible solutions in the time limit of 3600 s. The heuristic approach is based on a previous state of the art BVNS metaheuristic, which best configuration was able to reach the optimal value for the instances which the B&B certified the optimum, and better solutions than the B&B for larger instances. On the other hand, despite the fact that the studied heuristic found promising solutions for both variants of the problem, we observed significant differences in the values of the objective function.

Given the findings of this paper, specific heuristic algorithms for the MinSA should be designed when the host graph is a cycle. Furthermore, the relation of the two variants could be studied in terms of upper and lower bounds.

Appendix: Individual results per instance

Next, we present the individual results per instance obtained by each algorithm. In the case of the B&B procedure, results obtained before 3600 s (when the method is truncated) correspond to optimal solutions (Tables 3, 4 and 5).

Table 3. Complete instances.

Instance	BVNS$_1$		BVNS$_2$		B&B	
	F.O.	CPU T.(s)	F.O.	CPU T.(s)	F.O.	CPU T.(s)
Complete 002	24	30.00	10	30.00	10	4.82
Complete 004	26	30.00	9	30.01	9	6.62
Complete 006	4	30.00	4	30.00	4	3.33
Complete 009	2	30.00	2	30.00	2	3.26
Complete 012	523	30.10	464	30.42	654	3600.04
Complete 015	487	30.08	386	30.51	671	3600.04
Complete 018	511	30.10	406	30.24	654	3600.04
Complete 021	3127	30.47	2742	32.25	3912	3600.57
Complete 024	2974	30.35	2414	30.71	3789	3600.55
Complete 027	3011	30.18	2593	32.52	3958	3600.55
Complete 030	8752	31.15	8047	36.98	11431	3603.08
Complete 033	9248	31.49	8490	41.44	11677	3603.14
Complete 036	8665	30.23	7848	32.83	11044	3602.82
Complete 039	21155	31.93	18749	154.85	25953	3610.93
Complete 042	19375	32.40	17525	160.70	24799	3610.29
Complete 045	20418	31.87	18652	94.07	25443	3610.58
Complete 048	40977	34.45	36861	367.09	49215	3629.08
Complete 050	67360	33.26	64750	430.39	76242	3645.31
Complete 052	42203	35.66	38628	170.82	48215	3632.31
Complete 055	70728	39.95	66651	602.33	80189	3675.44

Table 4. Interval instances.

Instance	BVNS$_1$		BVNS$_2$		B&B	
	F.O.	CPU T.(s)	F.O.	CPU T.(s)	F.O.	CPU T.(s)
Interval 001	0	0.00	0	0.00	0	0.04
Interval 003	0	0.00	0	0.00	0	0.01
Interval 005	0	0.00	0	0.00	0	0.04
Interval 008	0	0.00	0	0.00	0	0.02
Interval 012	0	0.01	0	0.01	0	6.05
Interval 025	395	0.11	210	31.98	1802	3600.32
Interval 027	44	30.35	81	30.38	457	3600.30
Interval 029	10	30.68	7	31.57	179	3600.17
Interval 034	210	0.39	102	34.18	5070	3601.88
Interval 038	83	30.06	4	32.36	707	3600.66
Interval 043	136	0.63	80	64.96	12056	3606.49
Interval 048	16	30.55	27	36.10	648	3601.23
Interval 056	364	32.13	359	53.19	3013	3604.36
Interval 057	19	31.85	13	42.14	810	3603.17
Interval 060	609	31.49	440	87.84	9273	3616.54
Interval 065	325	33.33	304	37.73	5007	3609.50
Interval 074	790	34.03	617	82.09	5300	3616.70
Interval 075	122	32.46	96	62.86	3581	3612.18
Interval 082	749	32.19	394	103.82	10124	3627.70
Interval 084	256	33.27	221	142.77	3528	3620.45

Table 5. Random instances.

Instance	BVNS$_1$		BVNS$_2$		B&B	
	F.O.	CPU T.(s)	F.O.	CPU T.(s)	F.O.	CPU T.(s)
Random 001	0	0.00	0	0.00	0	0.05
Random 003	0	0.00	0	0.00	0	0.00
Random 005	1	30.00	0	0.00	0	0.01
Random 009	0	0.00	0	0.00	0	0.18
Random 014	155	30.10	119	30.43	224	3600.01
Random 018	241	30.03	223	30.59	395	3600.03
Random 021	18	30.36	9	30.64	100	3600.03
Random 025	2391	30.34	1733	31.76	2677	3600.44
Random 029	396	30.20	235	31.37	625	3600.18
Random 034	6111	30.79	4945	33.31	7216	3602.26
Random 038	845	30.36	616	33.17	1428	3600.72
Random 043	14527	37.21	12575	115.32	16561	3608.11
Random 049	8879	32.60	7011	93.09	11043	3608.58
Random 055	2068	32.19	1252	58.22	2854	3603.38
Random 059	25137	33.62	22905	313.72	30730	3629.00
Random 064	3577	31.92	2392	91.39	4711	3607.95
Random 069	21834	37.00	19804	570.10	29799	3637.83
Random 075	3644	39.16	2947	114.53	5874	3611.89
Random 080	161249	42.01	156263	7761.31	183124	3887.71
Random 086	85463	62.94	80143	3520.84	100308	3810.56

References

1. Aracena, J., Caro, C.T.: The weighted sitting closer to friends than enemies problem in the line. arXiv preprint arXiv:1906.11812 (2019)
2. Becerra, R., Caro, C.T.: On the sitting closer to friends than enemies problem in trees and an intersection model for strongly chordal graphs. arXiv preprint arXiv:1911.11494 (2019)
3. Benítez, F., Aracena, J., Caro, C.T.: The sitting closer to friends than enemies problem in the circumference. arXiv preprint arXiv:1811.02699 (2018)
4. Bhatt, S.N., Leighton, F.T.: A framework for solving VLSI graph layout problems. J. Comput. Syst. Sci. **28**(2), 300–343 (1984)
5. Cancho, R.F.: Euclidean distance between syntactically linked words. Phys. Rev. E **70**(5), 056135 (2004)
6. Cavero, S., Pardo, E.G., Duarte, A.: Efficient iterated greedy for the two-dimensional bandwidth minimization problem. Eur. J. Oper. Res. (2022). https://doi.org/10.1016/j.ejor.2022.09.004, in press
7. Cavero, S., Pardo, E.G., Duarte, A.: A general variable neighborhood search for the cyclic antibandwidth problem. In: Computational Optimization and Applications, pp. 1–31 (2022)
8. Cavero, S., Pardo, E.G., Duarte, A., Rodriguez-Tello, E.: A variable neighborhood search approach for cyclic bandwidth sum problem. Knowl. Based Syst. 108680 (2022)
9. Cavero, S., Pardo, E.G., Laguna, M., Duarte, A.: Multistart search for the cyclic cutwidth minimization problem. Comput. Oper. Res. **126**, 105116 (2021)
10. Civicioglu, P.: Backtracking search optimization algorithm for numerical optimization problems. Appl. Math. Comput. **219**(15), 8121–8144 (2013)
11. Cygan, M., Pilipczuk, M., Pilipczuk, M., Wojtaszczyk, J.O.: Sitting closer to friends than enemies, revisited. In: Rovan, B., Sassone, V., Widmayer, P. (eds.) MFCS 2012. LNCS, vol. 7464, pp. 296–307. Springer, Heidelberg (2012). https://doi.org/10.1007/978-3-642-32589-2_28
12. Díaz, J., Petit, J., Serna, M.: A survey of graph layout problems. ACM Comput. Surv. (CSUR) **34**(3), 313–356 (2002)
13. Ding, G., Oporowski, B.: Some results on tree decomposition of graphs. J. Graph Theory **20**(4), 481–499 (1995)
14. Duarte, A., Pantrigo, J.J., Pardo, E.G., Sánchez-Oro, J.: Parallel variable neighbourhood search strategies for the cutwidth minimization problem. IMA J. Manag. Math. **27**(1), 55–73 (2016)
15. Farber, M.: Characterizations of strongly chordal graphs. Discret. Math. **43**(2–3), 173–189 (1983)
16. Hansen, P., Mladenović, N.: Variable neighborhood search: principles and applications. Eur. J. Oper. Res. **130**(3), 449–467 (2001)
17. Hansen, P., Mladenović, N., Todosijević, R., Hanafi, S.: Variable neighborhood search: basics and variants. EURO J. Comput. Optim. **5**(3), 423–454 (2017)
18. Kermarrec, A.-M., Thraves, C.: Can everybody sit closer to their friends than their enemies? In: Murlak, F., Sankowski, P. (eds.) MFCS 2011. LNCS, vol. 6907, pp. 388–399. Springer, Heidelberg (2011). https://doi.org/10.1007/978-3-642-22993-0_36

19. Land, A.H., Doig, A.G.: An automatic method for solving discrete programming problems. In: Jünger, M., Liebling, T.M., Naddef, D., Nemhauser, G.L., Pulley-blank, W.R., Reinelt, G., Rinaldi, G., Wolsey, L.A. (eds.) 50 Years of Integer Programming 1958-2008, pp. 105–132. Springer, Heidelberg (2010). https://doi.org/10.1007/978-3-540-68279-0_5
20. Little, J.D., Murty, K.G., Sweeney, D.W., Karel, C.: An algorithm for the traveling salesman problem. Oper. Res. **11**(6), 972–989 (1963)
21. Pardo, E.G., García-Sánchez, A., Sevaux, M., Duarte, A.: Basic variable neighborhood search for the minimum sitting arrangement problem. J. Heuristics **26**(2), 249–268 (2020)
22. Pardo, E.G., Martí, R., Duarte, A.: Linear layout problems. In: Martí, R., Pardalos, P.M., Resende, M.G.C. (eds.) Handbook of Heuristics, pp. 1025–1049. Springer, Cham (2018). https://doi.org/10.1007/978-3-319-07124-4_45
23. Pardo, E.G., Soto, M., Thraves, C.: Embedding signed graphs in the line. J. Comb. Optim. **29**(2), 451–471 (2015)
24. Rodriguez-Tello, E., Hao, J.K., Torres-Jimenez, J.: An effective two-stage simulated annealing algorithm for the minimum linear arrangement problem. Comput. Oper. Res. **35**(10), 3331–3346 (2008)
25. Sánchez-Oro, J., Pantrigo, J.J., Duarte, A.: Combining intensification and diversification strategies in VNS an application to the vertex separation problem. Comput. Oper. Res. **52**, 209–219 (2014)
26. Tamassia, R.: Handbook of Graph Drawing and Visualization. CRC Press (2013)
27. Tucker, A.: An efficient test for circular-arc graphs. SIAM J. Comput. **9**(1), 1–24 (1980)
28. Wilcoxon, F.: Individual comparisons by ranking methods. In: In: Kotz, S., Johnson, N.L. (eds.) Breakthroughs in Statistics. Springer Series in Statistics, pp. 196–202. Springer, New York (1992). https://doi.org/10.1007/978-1-4612-4380-9_16

Assigning Multi-skill Configurations to Multiple Servers with a Reduced VNS

Thiago Alves de Queiroz[1]([envelope]) [iD], Beatrice Bolsi[2] [iD], Vinícius Loti de Lima[3] [iD], Manuel Iori[2] [iD], and Arthur Kramer[4,5] [iD]

[1] Institute of Mathematics and Technology, Federal University of Catalão, Catalão, GO 75704-020, Brazil
taq@ufcat.edu.br
[2] Department of Sciences and Methods for Engineering, University of Modena and Reggio Emilia, 42122 Reggio Emilia, Italy
{beatrice.bolsi,manuel.iori}@unimore.it
[3] Amazon.com, Seattle, USA
vloti@amazon.com
[4] Department of Production Engineering, Federal University of Rio Grande do Norte, Natal, RN 59077-080, Brazil
arthur.kramer@emse.fr
[5] Mines Saint-Etienne, Univ Clermont Auvergne, CNRS, UMR 6158 LIMOS, Henri Fayol Institute, Saint-Etienne 42023, France

Abstract. In this work, we deal with a dynamic problem arising from outpatient healthcare facility systems. Patients in need of service arrive during the day at the facility. Their requests are expected to be satisfied within a given target time, otherwise, tardiness is incurred. The facility has multiple identical servers that operate simultaneously and are in charge of providing the patients with the requested services. Each server can provide only a finite subset of services, and each subset is called a configuration. The objective is to assign to each server a configuration selected from a set of predefined configurations, aiming at minimizing total tardiness. Assignments are not fixed statically, but they can be dynamically changed over time to better cope with the requested services. As the problem nature is dynamic, we propose a re-optimization algorithm that periodically optimizes the assignments with a Reduced Variable Neighborhood Search (RVNS). The RVNS works on neighborhood structures based on changing the assignments of one or more servers. The RVNS has been extensively tested on realistic instances. The results prove its efficiency in reaching low-tardiness solutions under low computing time.

Keywords: Dynamic outpatient facility · Reduced VNS · Total tardiness

1 Introduction

Queue management is widely studied in many research fields and has both, high practical and theoretical relevance. Queue management problems are found

---

This work does not relate to the author's position at Amazon.

© The Author(s), under exclusive license to Springer Nature Switzerland AG 2023
A. Sleptchenko et al. (Eds.): ICVNS 2022, LNCS 13863, pp. 97–111, 2023.
https://doi.org/10.1007/978-3-031-34500-5_8

in several different contexts. One particular context is healthcare, where the presence of multiple queues and servers, dynamic arrival of patients, unknown service times, and different priorities, among many other features, provide several challenges [3, 22, 27]. General queue management problems have been addressed by a variety of approaches, in particular, as the focus of this paper, by operations research methods [25, 26].

The problem we address in this paper is motivated by a real-world situation faced by a healthcare facility. The problem concerns the queue management optimization in an outpatient facility system, where patients requiring specific services arrive dynamically during the working time horizon. For each patient arriving in the facility, the requested service and the expected service time are assumed to be known. The patients are grouped in multiple queues, one per service, and served by multiple identical servers working in parallel. At a given time instant, each server is able to provide a subset of all possible services provided by the facility. This subset is called *configuration* in the following. Thus, a patient can be served by a given server in a given time instant only if the server can provide the service requested by the patient, that is if the configuration currently adopted by the server includes the requested service.

During the considered time horizon, the configuration associated with each server is allowed to change. In practice, this possibility enables the facility to adapt the offered services according to the overall needs of the patients in the queues. The possibility of changing configurations during the facility's operating hours may help, for example, in reducing the patients' waiting time. The possible configurations the servers can assume belong to a finite set and are known in advance. In addition to the requested service and the expected service time, each patient is associated with a service-dependent target time. Tardiness is incurred if the time in which a patient is extracted by her queue is larger than her target time. The decision to be taken is which configuration to assign to each server over the time horizon so as to minimize the total tardiness of patients.

Many works in the literature deal with healthcare problems in Emergency Departments (ED) focusing on the scheduling of patients, nurses, operation rooms, surgeries, and others, with the aim of optimizing Key Performance Indicators (KPI) such as patients waiting time and length of stay [1]. Regarding queuing models in literature, there is a large variety of works addressing theoretical aspects. In [27], the authors reviewed the advances in queuing theory regarding applications in healthcare, focusing on queuing systems with infinite number of servers. Other authors studied problems with multiple servers and single/multiple queues, e.g., [11] who proposed an algorithm that adjusts the number of open servers depending on the number of patients in a single queue. Concerning practical applications, [13] considered a queuing system in a fast-food company composed of a finite number of servers and a capacitated queue. The authors focused on controlling the queue size by the allocation of workers to servers. Regarding the semiconductor industry, [12] studied a five-stage queuing system. In the healthcare context, [19] discussed real applications of queuing theory. In [23], the authors studied the registration process of a hospital located in China through simulation. In order to reduce the patients waiting time, the authors proposed to change from a multiple queue and multiple servers to a single queue with multiple servers. Still, with the objective of reducing patients waiting time, [24] investigated an ED in Romania. For a general discussion about queuing theory in healthcare, we refer the reader to [6].

We tackle the problem under investigation by proposing a re-optimization algorithm combined with a Reduced Variable Neighborhood Search (RVNS) metaheuristic to decide configuration-server assignments over a given time horizon. We also propose and test some variants of the RVNS. We aim at improving patient satisfaction by minimizing total tardiness. This work is a follow-up of [4], where we proposed some constructive heuristics.

The remainder of the paper is organized as follows. Section 2 contains the problem definition and its context. Section 3 is devoted to the re-optimization algorithm, the RVNS details and its variants. In Sect. 4, we present the numerical results, discussing the impact of the number of available servers and the way the patients are scheduled. Section 5 presents concluding remarks and directions for future investigations.

2 Problem Definition

The problem we handle comes from an existing situation experienced by an out-patient facility system in the north of Italy. The facility receives many patients per day and has multiple servers operating in parallel to serve these patients. At the facility, the patients first book a ticket for the service they need to receive and then stay in a waiting room until being called by one of the servers. In case two or more patients request the same service, they are inserted in the same First-In-First-Out (FIFO) queue associated with the service. As each patient has a target time, the facility should accomplish the requested service within this target time; otherwise, it will incur tardiness. We assume the servers are identical, with the same efficiency, and each one can serve one patient per time. Besides that, to each server is assigned a configuration (i.e., a subset of services that a server can serve). Due to the dynamic nature of the problem, the configuration of each server can change with time.

In other words, the problem considers a set S of identical parallel servers, a set K of services offered by the facility, a set C of possible configurations (from having no service to having all services in K), and a set I of patients. The sets S, K, and C are static, while the set I is dynamic, i.e., I changes over time, as patients arrive, receive their services, and leave the facility. Each server $s \in S$ can hold any configuration $c \in C$. Each configuration $c \in C$ represents a subset of services in K. Each service $k \in K$ has an expected processing time p_k, which is the average over the historical data, and holds a queue Q_k, with the patients who requested that service and are waiting to be served. Each patient $i \in I$ has an arrival time a_i (i.e., when she enters the queue), a service $k_i \in K$ (i.e., the service she requested – we assume each patient can only request a single service), and a target time d_i (i.e., the due time to be served).

There is a discrete-time horizon T (in minutes) over which the facility (servers) is operating and receiving (servicing) patients. The decisions are related to assigning one configuration in C to each server in S at each instant of the time horizon T, and to define the start time t_i at which each patient $i \in I$ is served by a server in S. The problem constraints are: all patients must receive their services; a patient can only be served if there is a server currently configured with her requested service; two or more servers can hold the same configuration; the configurations adopted by the servers can only be changed every Δ units of time (and after finishing the service of their current patients, if any); and each server can serve at most one patient per time. The problem objective is to minimize the total tardiness z, where the tardiness of a patient i is given by $T_i = \max\{0; \ t_i - d_i\}$. We summarize in Table 1 the parameters and variables used in this section.

Table 1. Parameters and variables used to describe the problem.

Variable	Description
T	End of the time horizon
I	Set of patients
S	Set of parallel servers
K	Set of available services
C	Set of possible configurations
p_k	Expected processing time of service $k \in K$
Q_k	Queue with the patients requiring service $k \in K$
a_i	Arrival (release) time of patient $i \in I$
k_i	Service requested by patient $i \in I$
d_i	Target (due) time of patient $i \in I$
t_i	Start time for patient $i \in I$
T_i	Tardiness generated after serving patient $i \in I$
z	Total tardiness (sum of the patient's tardiness) generated over the time horizon
N_T	Number of tardy patients over the time horizon
Δ	Time-step to change the servers configurations

We notice that the static version (i.e., when all data is known in advance) of this problem is related to the identical parallel machine scheduling problem with release dates and minimizing the total tardiness. For a more detailed mathematical description we refer the reader to [4]. This problem is NP-hard and well-studied in the scheduling literature [2,15,18]. As the problem we are facing is dynamic by nature, we opted for developing a fast and accurate re-optimization algorithm equipped with the RVNS, as described in the next section.

3   Proposed Algorithm

Variable Neighborhood Search (VNS) algorithms are based on the principle of systematically changing neighborhoods whenever the incumbent solution is not improved, to escape from valleys and attain a final solution that is globally optimal concerning all neighborhoods. VNS was proposed in [16] and since then it has been successfully applied to handle many different optimization problems, including integer linear and non-linear programming models. Comprehensive reviews dedicated to VNS, including recent advances, extensions, issues, and/or problem applications, are provided in, e.g., [7,9,10]. Furthermore, special issues dedicated to the VNS can be found in [5,17]. Recently, Lan et al. [14] reviewed the applications of VNS concerning problems in the healthcare area.

In a basic VNS, we have three distinct phases constituting the main algorithm schemes. The first is the *shaking* (i.e., a stochastic search, where a random solution is generated from a given initial solution and a given neighborhood). The second is the *local search* (i.e., a search based on the best/first improvement method) in the solution obtained from the previous phase. The last is related to the *change of neighborhood* (i.e., if the new solution improves the incumbent one, the incumbent solution is updated, and the search restarts from the first neighborhood structure). Another scheme is the general VNS, where the local search consists of the variable neighborhood descent method (i.e., a deterministic search over the neighborhoods). Both basic and general VNSs can use both stochastic and deterministic changes of neighborhoods [8].

In terms of computing time, one bottleneck of these schemes may be the local search, especially for very large instances or, in our case, dynamic problems where decisions should be taken in a very short time to avoid service interruptions. Therefore, we use the Reduced VNS (RVNS) for the problem under consideration. The RVNS consists of the first and last phases of the main scheme, i.e., it has the shaking and the change of neighborhood phases, as described in Algorithm 1. Its input parameters are a given solution x, the maximum number $It_{\max}$ of iterations used as stopping criterion, and the number $N_{\max}$ of neighborhood structures.

Algorithm 1: FRAMEWORK OF THE RVNS

Input: x; $It_{\max}$; $N_{\max}$.

```
1  for j ← 1, 2, . . . , It_max do
2  │   n ← 1
3  │   do
   │   │   /*shaking*/
4  │   │   x′ ← neighbor solution of x from structure N_n
   │   │   /*change of neighborhood*/
5  │   │   if f(x′) < f(x) then
6  │   │   │   x ← x′; n ← 1
7  │   │   else
8  │   │   └   n ← n + 1
9  │   while n ≤ N_max
```

Output: x

Due to the dynamic nature of the problem, where patients are arriving during the facility's working hours, we develop a re-optimization algorithm that uses the RVNS to update the solution in the occurrence of an event. Re-optimization (greedy) algorithms take decisions based on the currently available information. They usually do not explore the stochastic aspect of the problem [20]. We assume there is a single event, called *update servers*, occurring every Δ minutes. Therefore, we discretize the time horizon in slots of Δ minutes (i.e., $0, \Delta, 2\Delta, \ldots, T$), although patients can arrive at any (discrete) time, starting from zero. Algorithm 2 describes the proposed re-optimization algorithm, where each call to the RVNS takes into consideration the patients waiting for service in I and the current time j.

Algorithm 2: RE-OPTIMIZATION ALGORITHM WITH RVNS

   Input: $It_{\max}$; $N_{\max}$; Δ; T.

1   $x \leftarrow$ randomly generated solution

2   **for** $j \leftarrow 0, 1, 2, \ldots, T$ **do**

3      $I \leftarrow$ update the set of patients

4   **if** j *is a multiple of* Δ **then**

5         $x \leftarrow \text{RVNS}(x, It_{\max}, N_{\max})$

6   Schedule patients in I to servers in S, according to the configurations adopted in x

---

In Algorithm 2, the solution x is coded as a vector of elements. Each position of the vector is associated both with a server $s \in S$ and its current configuration $c \in C$, and an ordered list of patients scheduled to such server s. The initial solution x is created in the following way: for each server $s \in S$, select randomly a configuration in C and assign it to s. In the beginning, the list of patients served by s is empty since the arrival time of each patient cannot be less than the start of the time horizon. In the loop of lines 2–6, the set I of patients is updated. This means that patients already served and patients receiving a service at time j are *disregarded* from now on, while non-served yet patients and newly arrived ones can still be scheduled. This set, the time horizon T, and the current time j are global parameters used by the RVNS in Algorithm 1. Besides that, we propose a shaking phase using the following $N_{\max} = 3$ neighborhood structures:

- Neighborhood N_1: randomly select one server and change its configuration to a randomly selected different configuration in C. The new configuration of the server must be different from its current one.
- Neighborhood N_2: randomly select two servers and set the first server to have the same configuration as the second one;
- Neighborhood N_3: randomly select a subset of servers and change the configuration of each of these servers to a randomly selected configuration in C. The new configuration of each server may be the current one.

The call to the RVNS in line 5 of Algorithm 2 is responsible to update x by optimizing the servers' configuration. In line 6, we assume the *non-disregarded* patients in I are scheduled by the FIFO policy, i.e., the servers in the solution x have their ordered list of patients updated by the assignment of patients observing their arrival times (smallest first). We also investigate the influence of another policy in the computational experiments section.

The cost of a solution x is calculated with the function $f()$ in Algorithm 3, which considers the patients in I, the time horizon T, and the current time j. In this way, a solution x has its cost given by a first term, calculated over the disregarded patients in I, and a second term, calculated over the other patients in I. The value of the second term depends on the servers' configurations and how the non-disregarded patients in I are scheduled to the servers, from the current time j. Patients are scheduled by the FIFO policy and those requesting the same service are kept in the same queue. Therefore, if there is a free server whose configuration has the patient service, the patient with the smallest arrival time will start receiving service from such a server.

Algorithm 3: COST FUNCTION $f()$

Input: x; I; T; j.

1 $first_{cost} \leftarrow$ sum of the tardiness of the patients not in I
2 $second_{cost} \leftarrow 0$
3 **for** $t \leftarrow j, j+1, j+2, \ldots, T$ **do**
4      $S_{free} \leftarrow$ set of servers free at time t
5   **foreach** $s \in S_{free}$ **do**
6          $t_i \leftarrow$ schedule patient $i \in I$ with the smallest arrival time and whose service is offered by s
7          $second_{cost} \leftarrow$ update with the tardiness of patient i

Output: ($first_{cost}$+ $second_{cost}$)

In Algorithm 3, the free servers in S_{free} are those not servicing any patient at the current time t. In the loop of lines 5–7, patients are scheduled at t to the free servers. Patients can only be scheduled to a server if such a server is configured with that service at t. In the case a free server is not able to service any patient of I at t, it will continue free until there is a change in its configuration or new patients arrive. The $second_{cost}$ is updated with the tardiness of patient i that is scheduled to server s with start time t_i. Notice that scheduled patients of I cannot be considered in the next iterations of the algorithm.

In order to further explore the proposed neighborhood structures N_i, we implemented three additional versions of the RVNS, namely $RVNS_i$, for $i = 1, 2, 3$. To each $RVNS_i$, we associate three neighborhood structures N_i^k, for $k = 1, 2, 3$, in which N_i^1 is the same N_i. Besides that:

- $RVNS_1$ has, in this order, the neighborhood structures $\{N_1^k : k = 1, 2, 3\}$. In N_1^k, we perform a loop from $1, \ldots, k$, invoking the neighborhood N_1, that is N_1 is called k times;
- $RVNS_2$ has, in this order, the neighborhood structures $\{N_2^k : k = 1, 2, 3\}$. In N_2^k, we perform a loop from $1, \ldots, k$, invoking the neighborhood N_2, that is N_2 is called k times;

– RVNS$_3$ has, in this order, the neighborhood structures $\{N_3^k : k = 1, 2, 3\}$. In N_3^k, we invoke the neighborhood N_3 where the subset of servers has cardinality fixed to $k + 1$.

4   Computational Experiments

We coded all algorithms in the C++ language programming and tested them on 18 realistic instances, 15 of them already used by [4]. These instances were collected from an outpatient facility system in the north of Italy and corresponded to 18 working days in December 2019. The facility operates from 7 AM to 4 PM, with at most 13 servers serving between 147 to 1336 patients per day. The facility may offer 512 services organized in 128 different configurations. Moreover, as explained in [4], the facility prefers to update the servers' configuration every $\Delta = 60$ min. The experiments that we performed were carried out on a computer equipped with an Intel Core i7 processor of 1.2 GHz, 8 GB of RAM, and running macOS 12.2.1.

The parameters of the re-optimization algorithm were calibrated in a trial-and-error procedure, balancing solution quality and computing time. After the calibration, we set $It_{\max} = 250$ iterations and for the neighborhood N_3, the size of the subsets generated are in the set $\{2, 3, \ldots, \lfloor |S|/4 \rfloor\}$. In the following, we discuss how the problem parameters influence the solution quality, by focusing in particular on the number $|S|$ of available servers, the value of Δ, and the policy used to schedule patients. The solution quality is measured in terms of the total tardiness z (i.e., the main problem objective) and the number of tardy patients N_T (an additional interesting KPI). The algorithm was executed 10 times with different seeds, and average values are reported per instance. The computing times are not reported because they were always below 3 s per execution. Despite the $|S| = 13$ servers available in the facility, keeping all of them operating may be too costly. Similarly, the facility would like to avoid changing the servers' configuration constantly, and the value of Δ influences the number of times the RVNS is applied to optimize (and possibly change) the servers' configurations. Concerning the policy to schedule patients, besides the FIFO, we also implemented the Earliest Due Date (EDD) rule. In the EDD, the patients are scheduled by the smallest target time.

The first results we present are related to a comparison between the re-optimization algorithm with RVNS and the re-optimization algorithm with the other RVNS variants (i.e., RVNS$_1$, RVNS$_2$, and RVNS$_3$). In these tests, we set $\Delta = 60$ and $|S| = 13$, besides considering the FIFO and EDD policies. We compare the performance of the four algorithms. Table 2 has the results by using the FIFO policy, while Table 3 has the results by using the EDD policy. Each line of these tables presents the instance number, the number of patients in I, and, for each algorithm, the total tardiness z and the number of tardy patients N_T.

In the results of Tables 2 and 3, we observe that the RVNS$_2$ could not achieve a feasible solution for all the executions, which are the average values marked with an *. In this way, we do not consider this algorithm when comparing it with the other ones. For the FIFO policy, in Table 2, we observe the RVNS, as originally proposed, has the best overall results in terms of total tardiness and number of tardy patients. It presents the best (or equal) average total tardiness for 12 out of 18 instances, while the $RVNS_1$ and $RVNS_3$ have the best (or equal) results for 5 and 9 instances, respectively. On the other hand, considering the EDD policy, in Table 3, the $RVNS_1$ has the best overall results in terms of total tardiness, while the $RVNS_3$ has the best overall results in terms of the number of tardy patients. Considering the results for each instance, the RVNS, $RVNS_1$, and $RVNS_3$ have the best (or equal) average total tardiness for 6, 11, and 7 instances, respectively. After all, we have decided to use the re-optimization algorithm with RVNS in the next experiments.

Table 2. Results for different RVNS algorithms, using the FIFO policy.

Inst.	$|I|$	RVNS		$RVNS_1$		$RVNS_2$		$RVNS_3$	
		z	N_T	z	N_T	z	N_T	z	N_T
1	147	0.00	0.00	0.00	0.00	0.00	0.00	0.00	0.00
2	215	0.00	0.00	0.00	0.00	0.00	0.00	0.00	0.00
3	914	293.11	64.50	284.94	64.40	272.09*	63.00	283.75	64.40
4	959	64.63	20.10	96.94	21.30	25.53*	17.70	67.06	19.40
5	974	104.98	32.20	84.08	31.10	83.18*	31.00	84.08	31.10
6	1034	308.56	66.10	309.56	65.80	307.70*	65.00	308.56	66.10
7	1052	176.25	54.00	176.25	54.00	176.25*	54.00	176.25	54.00
8	1052	110.74	52.20	143.25	53.70	110.91*	52.30	186.56	56.00
9	1105	163.34	36.90	147.09	35.90	115.32*	33.60	142.36	35.60
10	1123	180.13	48.30	175.11	47.80	166.52*	46.60	172.15	47.20
11	1124	346.69	65.00	356.60	65.80	301.14*	60.60	355.77	65.10
12	1160	179.83	56.90	197.15	58.20	93.68*	51.40	270.35	61.40
13	1193	101.19	47.50	87.86	46.50	71.44*	44.20	109.73	47.10
14	1193	167.46	48.40	171.01	48.00	153.43*	45.80	168.43	47.00
15	1217	897.87	253.30	900.78	253.10	829.07*	247.80	896.84	251.90
16	1276	369.57	95.20	371.98	95.30	352.25*	93.10	375.91	96.70
17	1309	401.27	80.90	471.61	85.50	353.52*	74.50	428.01	84.00
18	1336	243.98	116.50	251.97	114.10	221.61*	112.80	250.47	117.10
Average		228.31	63.22	234.79	63.36	201.87*	60.74	237.57	63.56

Table 4 has the results that we obtained for different number of servers, i.e., $|S| \in \{4, 6, 8, 10, 13\}$, and $\Delta = 60$, using the FIFO policy to schedule patients. The results obtained with the EDD policy are presented in Table 5. Observing the results in these tables, we notice that increasing the number of available servers is beneficial to improve the solution quality. By increasing from 4 to 13 servers, we can reduce the average total tardiness and the average number of tardy patients by 99.88% and 93.64%, with the FIFO policy, and 99.90% and

Table 3. Results for different RVNS algorithms, using the EDD policy.

Inst.	$|I|$	RVNS		RVNS$_1$		RVNS$_2$		RVNS$_3$	
		z	N_T	z	N_T	z	N_T	z	N_T
1	147	0.00	0.00	0.00	0.00	0.00	0.00	0.00	0.00
2	215	0.00	0.00	0.00	0.00	0.00	0.00	0.00	0.00
3	914	275.64	57.10	270.80	56.50	266.68*	56.00	275.31	57.10
4	959	103.18	18.40	85.16	18.80	27.92*	15.40	104.53	18.60
5	974	89.75	32.50	89.75	32.50	84.17*	31.30	93.43	33.10
6	1034	302.43	59.20	301.12	59.10	301.53*	59.40	301.41	59.40
7	1052	170.67	46.20	170.96	46.20	170.65*	46.00	170.67	46.20
8	1052	173.61	32.80	123.29	31.00	87.11*	29.30	151.54	31.80
9	1105	148.78	36.00	141.15	35.70	121.49*	34.70	159.19	37.00
10	1123	163.25	41.60	162.14	42.90	154.09*	40.40	161.17	41.80
11	1124	401.31	68.00	363.08	65.00	299.80*	60.00	340.98	63.50
12	1160	147.64	39.80	184.90	41.60	78.13*	35.30	170.83	40.60
13	1193	89.21	32.70	86.54	31.80	59.72*	29.10	81.66	31.70
14	1193	170.85	45.60	173.74	46.10	150.14*	42.80	162.59	44.40
15	1217	379.15	80.10	370.29	78.20	347.51*	68.10	377.49	78.80
16	1276	326.15	71.00	320.01	70.70	298.04*	67.10	333.33	71.50
17	1309	407.39	76.80	430.07	77.10	357.29*	72.90	428.58	77.10
18	1336	149.52	39.10	139.41	39.80	119.55*	38.90	139.48	38.20
Average		194.36	43.16	189.58	42.94	162.43*	40.37	191.79	42.82

95.66%, with the EDD policy, respectively. Comparing the two policies used to schedule patients, EDD is relatively better, with an average overall reduction of 0.58%, in terms of total tardiness, and 2.70%, in terms of number of tardy patients, compared to FIFO.

Table 6 gives the results for different values of $\Delta \in \{15, 30, 60, 90, 120\}$, and $|S| = 13$ servers, using the FIFO policy. For the EDD policy, the results are shown in Table 7. We observe that smaller values of Δ lead to better quality solutions. In general, we can observe an average reduction of 98.92%, in terms of total tardiness, and 89.84%, in terms of the number of tardy patients, for the FIFO policy, when passing from $\Delta = 120$ to $\Delta = 15$. Even better results are obtained with the EDD policy, with an average reduction of 99.62% and 95.86%, respectively. Concerning the facility preference to update the servers every $\Delta = 60$ min, we notice that the solution quality could be improved if smaller values of Δ were considered, as the best results are indeed obtained with $\Delta = 15$. However, the decision maker should evaluate the managerial impact of these more frequent changes in the daily practice. Comparing the two policies, again EDD is better than FIFO, showing an overall average reduction of 5.86%, in terms of total tardiness, and 10.12%, in terms of the number of tardy patients with respect to FIFO.

Table 4. Results for different number of servers, using the FIFO policy.

Inst.	$|I|$	$|S| = 4$		$|S| = 6$		$|S| = 8$		$|S| = 10$		$|S| = 13$	
		z	N_T	z	N_T	z	N_T	z	N_T	z	N_T
1	147	0.00	0.00	0.00	0.00	0.00	0.00	0.00	0.00	0.00	0.00
2	215	3.85	5.00	0.00	0.00	0.00	0.00	0.00	0.00	0.00	0.00
3	914	118684.30	909.90	21044.50	691.50	1561.13	270.40	488.54	99.40	296.71	64.90
4	959	116877.20	948.00	13996.09	663.20	358.90	119.00	137.26	39.60	74.74	20.60
5	974	121110.80	966.40	14245.33	608.10	727.57	176.00	167.21	53.20	86.68	31.40
6	1034	173742.00	1030.00	38670.35	918.40	2926.80	382.00	597.02	124.10	308.25	65.70
7	1052	172469.30	1043.00	40537.99	996.10	2866.27	454.30	338.01	103.50	187.26	54.50
8	1052	166273.50	1048.00	30581.86	908.90	1774.57	343.80	293.97	79.70	165.92	54.80
9	1105	195975.60	1101.00	41352.55	1061.10	1566.48	299.60	276.81	84.90	149.24	35.60
10	1123	211362.60	1118.60	53127.21	1072.90	5741.46	601.80	327.82	83.80	174.83	47.80
11	1124	216109.50	1120.00	54217.47	1080.50	4956.95	531.90	744.29	133.80	381.49	67.30
12	1160	229894.70	1156.00	60780.00	1119.00	8788.97	621.70	712.12	198.10	199.31	58.20
13	1193	247645.20	1188.70	66317.00	1172.00	4187.31	579.90	303.52	120.30	100.55	47.30
14	1193	249619.70	1189.00	69532.00	1156.00	6560.83	659.00	362.82	112.70	177.51	48.80
15	1217	302848.50	1213.00	117030.60	1211.00	32296.72	979.00	6254.42	581.90	895.40	253.10
16	1276	314534.40	1271.00	107492.40	1266.00	15863.21	878.50	1147.74	284.70	393.28	98.30
17	1309	333459.20	1305.00	116179.00	1303.00	16034.79	910.90	1013.86	209.80	363.11	76.10
18	1336	381722.90	1332.00	156984.20	1327.00	47227.49	1163.70	8933.06	892.00	243.89	116.40
Average		197351.85	996.92	55671.59	919.71	8524.41	498.42	1227.69	177.86	233.23	63.38

Table 5. Results for different number of servers, using the EDD policy.

Inst.	$|I|$	$|S| = 4$		$|S| = 6$		$|S| = 8$		$|S| = 10$		$|S| = 13$	
		z	N_T	z	N_T	z	N_T	z	N_T	z	N_T
1	147	0.00	0.00	0.00	0.00	0.00	0.00	0.00	0.00	0.00	0.00
2	215	0.00	0.00	0.00	0.00	0.00	0.00	0.00	0.00	0.00	0.00
3	914	118356.20	910.00	20765.74	741.70	942.30	193.50	453.91	68.70	278.06	57.40
4	959	116690.40	955.00	13570.49	664.00	269.24	34.90	115.51	23.00	96.82	18.30
5	974	120867.70	970.00	13868.26	604.40	261.02	51.70	154.74	38.00	111.73	33.90
6	1034	173377.50	1030.00	38273.48	963.10	2094.74	428.50	541.62	75.80	301.53	59.40
7	1052	172205.70	1048.00	40196.36	1025.40	1540.03	388.70	276.00	60.00	170.71	46.60
8	1052	165994.60	1048.00	30045.79	968.70	813.42	260.10	292.94	46.80	141.15	31.40
9	1105	195642.70	1101.00	40995.50	1087.00	599.69	153.50	267.77	49.60	152.27	36.50
10	1123	210641.50	1119.00	52603.30	1090.00	5098.18	564.90	268.13	49.20	162.18	42.10
11	1124	215733.70	1120.00	53777.90	1118.00	4541.32	479.10	599.47	78.50	358.32	64.90
12	1160	229427.90	1156.00	60438.30	1135.00	8259.40	656.20	285.06	74.80	203.43	42.50
13	1193	247028.70	1189.00	66066.40	1187.00	3089.97	549.90	189.62	57.20	89.43	32.70
14	1193	249015.20	1189.00	68810.10	1167.00	5693.25	666.50	308.42	57.90	182.91	47.00
15	1217	302490.30	1213.00	116828.80	1211.00	31880.78	1002.20	5511.65	585.90	379.32	80.50
16	1276	314271.30	1272.00	107327.30	1270.00	15381.76	904.10	583.80	131.90	320.59	68.80
17	1309	333646.30	1305.00	115899.50	1303.00	15042.97	973.70	751.92	94.70	406.95	76.80
18	1336	381082.40	1332.00	156311.10	1330.00	46665.82	1191.80	8162.23	928.30	140.01	41.00
Average		197026.23	997.61	55321.02	936.96	7898.55	472.18	1042.38	134.46	194.19	43.32

Table 6. Results for different values of Δ, using the FIFO policy.

Inst.	$|I|$	$\Delta = 15$		$\Delta = 30$		$\Delta = 60$		$\Delta = 90$		$\Delta = 120$	
		z	N_T	z	N_T	z	N_T	z	N_T	z	N_T
1	147	0.00	0.00	0.00	0.00	0.00	0.00	0.00	0.00	0.00	0.00
2	215	0.00	0.00	0.00	0.00	0.00	0.00	0.00	0.00	0.37	1.00
3	914	30.97	23.10	44.26	24.60	293.11	64.50	1036.96	122.10	3726.76	274.80
4	959	1.70	3.20	9.63	5.00	64.63	20.10	588.67	69.00	2398.85	217.00
5	974	12.02	9.00	14.62	9.30	104.98	32.20	433.17	68.90	1626.03	183.30
6	1034	44.19	33.70	47.16	35.20	308.56	66.10	1231.23	136.50	5483.58	388.00
7	1052	36.75	29.00	36.75	29.00	176.25	54.00	438.79	85.00	3464.25	286.60
8	1052	31.37	28.40	38.59	29.50	110.74	52.20	790.59	114.50	3934.38	281.50
9	1105	19.66	11.60	22.24	12.10	163.34	36.90	745.95	111.90	4591.25	327.90
10	1123	36.89	22.10	43.84	22.20	180.13	48.30	783.66	105.20	3605.84	301.50
11	1124	50.51	24.30	51.31	24.30	346.69	65.00	1140.43	125.70	3926.11	271.10
12	1160	34.92	35.00	40.25	35.90	179.83	56.90	941.88	172.20	7061.14	413.10
13	1193	16.50	18.70	16.03	18.30	101.19	47.50	1207.36	151.70	5761.23	371.80
14	1193	20.33	16.10	21.08	16.30	167.46	48.40	814.00	109.20	5024.45	364.10
15	1217	625.60	223.20	617.26	222.60	897.87	253.30	6293.37	511.10	20858.07	734.20
16	1276	94.54	56.50	96.77	57.20	369.57	95.20	2331.29	237.10	10366.98	519.70
17	1309	26.26	17.00	37.95	18.40	401.27	80.90	2145.36	217.00	9129.37	518.10
18	1336	127.13	89.10	139.66	88.30	243.98	116.50	3993.83	400.70	21393.43	844.80
Average		67.18	35.56	70.97	36.01	228.31	63.22	1384.25	152.10	6241.78	349.92

Table 7. Results for different values of Δ, using the EDD policy.

Inst.	$|I|$	$\Delta = 15$		$\Delta = 30$		$\Delta = 60$		$\Delta = 90$		$\Delta = 120$	
		z	N_T	z	N_T	z	N_T	z	N_T	z	N_T
1	147	0.00	0.00	0.00	0.00	0.00	0.00	0.00	0.00	0.00	0.00
2	215	0.00	0.00	0.00	0.00	0.00	0.00	0.00	0.00	0.00	0.00
3	914	25.48	16.00	34.15	17.10	275.64	57.10	970.54	109.10	3739.17	287.50
4	959	0.20	0.40	6.94	1.90	103.18	18.40	522.32	62.10	2212.91	220.60
5	974	12.02	9.00	15.52	9.60	89.75	32.50	418.35	67.90	1507.96	168.40
6	1034	35.49	26.40	40.85	28.50	302.43	59.20	1125.84	128.10	5296.02	389.10
7	1052	31.16	21.10	33.90	21.50	170.67	46.20	479.24	79.50	3258.92	277.70
8	1052	6.90	5.00	10.69	5.70	173.61	32.80	764.94	113.90	3679.86	288.70
9	1105	19.96	11.70	23.44	12.60	148.78	36.00	722.85	105.50	4460.73	355.00
10	1123	23.51	15.50	31.66	16.40	163.25	41.60	629.19	87.20	3390.68	298.30
11	1124	50.36	23.30	51.91	24.30	401.31	68.00	1111.06	124.30	3913.82	259.00
12	1160	23.43	20.00	33.52	21.80	147.64	39.80	754.25	121.60	7020.14	387.90
13	1193	2.68	3.50	8.65	4.40	89.21	32.70	1044.60	129.30	5778.56	373.90
14	1193	18.93	13.00	16.76	12.30	170.85	45.60	772.24	102.20	4669.85	348.10
15	1217	82.73	43.90	100.46	48.00	379.15	80.10	6072.47	519.20	20706.81	734.60
16	1276	40.27	30.10	40.88	30.20	326.15	71.00	2076.62	260.30	10109.94	520.10
17	1309	24.41	15.20	54.12	19.20	407.39	76.80	1999.52	228.60	8727.51	508.70
18	1336	11.20	7.20	29.09	9.30	149.52	39.10	3412.88	434.80	19639.72	890.50
Average		22.71	14.52	29.58	15.71	194.36	43.16	1270.94	148.53	6006.25	350.45

5 Conclusions and Future Research

Optimizing the flow of patients in outpatient facilities is always a hard task. While the patients would like to receive their services as soon as possible, the facility could be also interested in reducing its costs and operating with a reduced number of servers/operators. In this paper, we study how the right configuration of servers can help at reducing patients' tardiness in a dynamic environment. The need for sophisticated and fast methods is essential for that aim. For this reason, we developed a re-optimization algorithm, which uses an RVNS to decide on the assignments of configurations to servers, in order to solve realistic instances.

In the computational experiments, we evaluated how the number of servers, the time interval to update the servers' configuration, and the policy used to schedule patients impact the solution quality. Using all 13 servers is the best decision, with an average reduction in total tardiness of 81.01%, for the FIFO policy, and 81.37%, for the EDD policy, compared to using only 10 servers. Of course, the more the number of available servers, the less the number of patients without tardiness; despite that, [4] noticed that defining correctly the set of configurations that the servers could offer can also impact the solution quality. Concerning the time interval to update the servers' configuration, we observe that large values of Δ can affect the solution quality and thus a decision-maker should pay attention and act as soon as she detects an increase in the tardiness. With respect to the scheduling of patients, it is not a wise decision, though very popular, to prefer their arrival time (FIFO policy) to their target time (EDD policy), as this leads indeed to an overall average increase of 0.74% in the total tardiness and 4.31% in the number of tardy patients.

We observe that there is room for improvement in terms of solution methods. A direction could consider the proposal of more elaborated heuristics for the scheduling of patients instead of using the FIFO or EDD policies. Another direction could consider methods based on sampled scenarios to anticipate future decisions. For this line of research, it would be interesting to take into consideration information about the priority/urgency and the realized processing time of services for scheduling the patients. We could also investigate the problem under a multi-objective perspective by proposing multi-objective VNS algorithms (e.g., as in [21]).

Acknowledgements. The authors would like to thank the support given by the National Council for Scientific and Technological Development (CNPq grants numbers 405369/2021-2 and 311185/2020-7), and the State of Goiás Research Foundation (FAPEG). We also thank MAPS Group S.p.A. for sharing relevant information and knowledge that helped develop this research.

References

1. Abdalkareem, Z.A., Amir, A., Al-Betar, M.A., Ekhan, P., Hammouri, A.I.: Healthcare scheduling in optimization context: a review. Heal. Technol. **11**, 445–469 (2021)

2. Allahverdi, A.: The third comprehensive survey on scheduling problems with setup times/costs. Eur. J. Oper. Res. **246**(2), 345–378 (2015)
3. Amir Elalouf, G.W.: Queueing problems in emergency departments: a review of practical approaches and research methodologies. Oper. Res. Forum **3**(1), 2 (2022)
4. Bolsi, B., Kramer, A., de Queiroz, T.A., Iori, M.: Optimizing a dynamic outpatient facility system with multiple servers. In: Amorosi, L., Dell'Olmo, P., Lari, I. (eds.) Optimization in Artificial Intelligence and Data Sciences. AIROSS, vol. 8, pp. 247–258. Springer, Cham (2022). https://doi.org/10.1007/978-3-030-95380-5_22
5. Duarte, A., Pardo, E.G.: Special issue on recent innovations in variable neighborhood search. J. Heuristics **26**, 335–338 (2020)
6. Green, L.: Queueing analysis in healthcare. In: Hall, R.W. (ed.) Patient Flow: Reducing Delay in Healthcare Delivery. ISOR, pp. 281–307. Springer, Boston (2006). https://doi.org/10.1007/978-0-387-33636-7_10
7. Hansen, P., Mladenović, N.: Variable neighborhood search: principles and applications. Eur. J. Oper. Res. **130**(3), 449–467 (2001)
8. Hansen, P., Mladenović, N., Brimberg, J., Pérez, J.A.M.: Variable neighborhood search. In: Gendreau, M., Potvin, J.-Y. (eds.) Handbook of Metaheuristics. ISORMS, vol. 272, pp. 57–97. Springer, Cham (2019). https://doi.org/10.1007/978-3-319-91086-4_3
9. Hansen, P., Mladenović, N., Moreno Pérez, J.A.: Variable neighbourhood search: methods and applications. Ann. Oper. Res. **175**(1), 367–407 (2010)
10. Hansen, P., Mladenović, N., Todosijević, R., Hanafi, S.: Variable neighborhood search: basics and variants. EURO J. Comput. Optim. **5**, 423–454 (2017)
11. Kaboudan, M.A.: A dynamic-server queuing simulation. Comput. Oper. Res. **25**(6), 431–439 (1998)
12. Kumar, S., Omar, M.K.: Stochastic re-entrant line modeling for an environment stress testing in a semiconductor assembly industry. Appl. Math. Comput. **173**(1), 603–615 (2006)
13. Lan, C.H., Lan, T.S., Chen, M.S.: Optimal human resource allocation with finite servers and queuing capacity. Int. J. Comput. Appl. Technol. **24**(3), 156–160 (2005)
14. Lan, S., Fan, W., Yang, S., Pardalos, P.M., Mladenovic, N.: A survey on the applications of variable neighborhood search algorithm in healthcare management. Ann. Math. Artif. Intell. **89**, 741–775 (2021)
15. Lenstra, J., Kan, A.R., Brucker, P.: Complexity of machine scheduling problems. In: Hammer, P., Johnson, E., Korte, B., Nemhauser, G. (eds.) Studies in Integer Programming, Annals of Discrete Mathematics, vol. 1, pp. 343–362. Elsevier (1977)
16. Mladenović, N., Hansen, P.: Variable neighborhood search. Comput. Oper. Res. **24**(11), 1097–1100 (1997)
17. Mladenović, N., Souza, M., Sörensen, K.: Special issue on "advances in variable neighborhood search". Int. Trans. Oper. Res. **25**(1), 427–427 (2018)
18. Mokotoff, E.: Parallel machine scheduling problems: a survey. Asia-Pac. J. Oper. Res. **18**(2), 193–242 (2001)
19. Peter, P.O., Sivasamy, R.: Queueing theory techniques and its real applications to health care systems - outpatient visits. Int. J. Healthc. Manag. **14**(1), 114–122 (2021)
20. Alves de Queiroz, T., Iori, M., Kramer, A., Kuo, Y.-H.: Scheduling of patients in emergency departments with a variable neighborhood search. In: Mladenovic, N., Sleptchenko, A., Sifaleras, A., Omar, M. (eds.) ICVNS 2021. LNCS, vol. 12559, pp. 138–151. Springer, Cham (2021). https://doi.org/10.1007/978-3-030-69625-2_11

21. Queiroz, T.A., Mundim, L.R.: Multiobjective pseudo-variable neighborhood descent for a bicriteria parallel machine scheduling problem with setup time. Int. Trans. Oper. Res. **27**(3), 1478–1500 (2020)
22. Rais, A., Viana, A.: Operations research in healthcare: a survey. Int. Trans. Oper. Res. **18**(1), 1–31 (2011)
23. Su, Q., Yao, X., Su, P., Shi, J., Zhu, Y., Xue, L.: Hospital registration process reengineering using simulation method. J. Healthc. Eng. **1**(1), 67–82 (2010)
24. Vass, H., Szabo, Z.K.: Application of queuing model to patient flow in emergency department. Case study. Procedia Econ. Finan. **32**, 479–487 (2015)
25. Weiss, E.N., Tucker, C.: Queue management: elimination, expectation, and enhancement. Bus. Horiz. **61**(5), 671–678 (2018)
26. Worthington, D.: Reflections on queue modelling from the last 50 years. J. Oper. Res. Soc. **60**(sup1), S83–S92 (2009)
27. Worthington, D., Utley, M., Suen, D.: Infinite-server queueing models of demand in healthcare: a review of applications and ideas for further work. J. Oper. Res. Soc. **71**(8), 1145–1160 (2020)

Multi-Round Influence Maximization: A Variable Neighborhood Search Approach

Isaac Lozano-Osorio$^{(\boxtimes)}$ [iD], Jesús Sánchez-Oro [iD], and Abraham Duarte [iD]

Universidad Rey Juan Carlos, Madrid, Spain
{isaac.lozano,jesus.sanchezoro,abraham.duarte}@urjc.es

Abstract. The study of Social Network Influence has attracted the interest of scientists. The wide variety of real-world applications of this area, such as viral marketing and disease analysis, highlights the relevance of designing an algorithm capable of solving the problem efficiently. This paper studies the Multiple Round Influence Maximization (MRIM) problem, in which influence is propagated in multiple rounds independently from possibly different seed sets. This problem has two variants: the non-adaptive MRIM, in which the advertiser needs to determine the seed sets for all rounds at the beginning, and the adaptive MRIM, in which the advertiser can select the seed sets adaptively based on the propagation results in the previous rounds. The main difficulty of this optimization problem lies in the computational effort required to evaluate a solution. Since each node is infected with a certain probability, the value of the objective function must be calculated through an influence diffusion model, which results in a computationally complex process. For this purpose, a metaheuristic algorithm based on Variable Neighborhood Search is proposed with the aim of providing high-quality solutions, being competitive with the state of the art.

Keywords: Information systems · Social networks · Influence maximization · Network science · Viral marketing · VNS

1 Introduction

Today, people manage multiple Social Networks (SNs) from different social positions, for example: the internet, information, propagation of ideas, social bond dynamics, disease propagation, viral marketing, or advertisement, among others [4,6,15,16].

SNs are exponentially increasing the number of active users. This growth is extended to the amount of behavioral data, and, therefore, all classical network-related problems are becoming computationally harder.

It is worth mentioning that SNs are used not only to spread positive information, but also malicious information. In general, research devoted to maximizing the influence of positive ideas is called Influence Maximization [26]. Thus, solving successfully this problem allows the decision-maker to decide the best way to

A. Sleptchenko et al. (Eds.): ICVNS 2022, LNCS 13863, pp. 112–124, 2023.
https://doi.org/10.1007/978-3-031-34500-5_9

propagate information about products and/or services. On the contrary, SNs can be also used for the diffusion of malicious information like derogatory rumors, disinformation, hate speech, or fake news. These examples motivate research on how to reduce the influence of negative information. This family of problems are usually know as Influence Minimization [14,23].

Researchers usually model an SN as a graph $G(V, E)$ where the set of nodes V represents the users and each relation between two users is modeled as a pair $(u, v) \in E$, with $u, v \in V$ indicating that user u is connected to or can even transmit information to user v. Kempe [12] originally formalized the influence model to analyze how information is transmitted among SN users. Given an SN with $|V| = n$ nodes where the edges (relational links) represent the spreading or propagation process on that network, the task is to choose a seed set nodes S of size $l < n$ with the aim of maximizing the number of nodes in the network that are influenced by the seed set S.

The evaluation of the influence [17,27] of a given seed set S requires the definition of an Influence Diffusion Model (IDM) [12]. This model is responsible for deciding which nodes are affected by the information received from their neighboring nodes in the SN. The most extended IDMs are: Independent Cascade Model (ICM), Weighted Cascade Model (WCM), Linear Threshold Model (LTM), and Triggering Model (TM). All of them are based on assigning an influence probability to each relational link in the SN. ICM, which is one of the most widely used IDMs, considers that the probability of influence is the same for each link and is usually a small probability, being 1% a widely accepted value. On the contrary, WCM considers that the probability that a user v will be influenced by an user u is proportional to the in-degree of user v, i.e., the number of users that can eventually influence the user v. Therefore, the probability of influencing the user v is defined as $1/d_{in}(v)$, where $d_{in}(v)$ is the in-degree of user v. LTM, requires a specific activation weight for each link in the SN. Given these weights, a user will be influenced if and only if the sum of the weights of its neighbors is larger than or equal to a given threshold.

This work considers the Triggering Model as IDM, as it is the IDM used in the best algorithm found in the literature [33], with the aim of providing a fair comparison. The TM is a generalization of ICM and LT where every node v independently chooses a random trigger set according to some distribution over subsets of its neighbors and is influenced if any of the nodes in its trigger sets are influenced. Note that the ICM model is a special case of TM where every edge $(u, v) \in E$ is associated with a probability $p_{u,v} \in [0, 1]$ and is set to zero if $p_{u,v} \notin [0, 1]$. The trigger set for each user v is selected in each round, conformed with those users u whose probability of directly influencing v is larger than or equal to p_{uv}, being this probability selected at random for each round.

A solution for the Multi-Round Influence Maximization problem (MRIM) consist of selecting R seed sets of size l, one for each round, i.e., $S = \{S_1, S_2, \ldots, S_R\}$. Notice that, since l nodes can be selected for each round, the total number of nodes conforming the final seed set S is equal to $l \cdot R$. The aim of MRIM is to maximize the number of active nodes following a specific IDM. The objective function value is then evaluated as:

$$MRIM(S) = IDM(S_1 \cup S_2 \cup \ldots \cup S_R)$$

where *IDM* represents the influence diffusion model considered. Then, the objective of the MRIM is to find the seed set for each round that maximizes the value of the objective function value. In mathematical terms,

$$S^\star \leftarrow \underset{S \in \mathbb{S}}{\arg \max} MRIM(S)$$

where $\mathbb{S}$ is the set of all possible combinations of seed sets for the problem under consideration.

Notice that, in the case in which a single round is considered, i.e., $R = 1$, the problem is equivalent to the well-studied Social Network Influence Maximization Problem (SNIMP). The classical SNIMP is $\mathcal{NP}$-hard [13], so both the non-adaptive and adaptive versions of MRIM are also $\mathcal{NP}$-hard.

As it was aforementioned, there are two different approximations with the MRIM problem. On the one hand, the non-adaptive approach consisting of selecting a number l of nodes per round without having information about the influenced users in each round. On the other hand, in the adaptive approach, users are influenced after the selection of each round is known, increasing the information available for the next rounds.

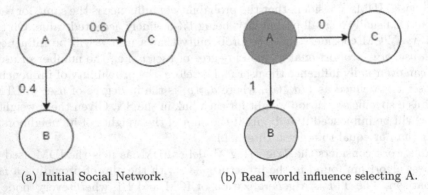

(a) Initial Social Network.   (b) Real world influence selecting A.

Fig. 1. Initial Social Network and influence propagation when A is considered as first seed set node.

Figure 1 represents an SN with 3 nodes and 2 relations where the number of selected nodes as a seed set is $l = 2$. Let us suppose that the initial selected node is always A. If we now consider the non-adaptive model, it is expected that the next most promising node to be selected is node C, since node C will be more probably influenced by A. However, if in the final simulation C is not finally influenced, the active nodes will be A and B. However, considering the adaptive model, after selecting the first node A, the method will know that C is not finally influenced, so it can select C to increase the number of active nodes, resulting in

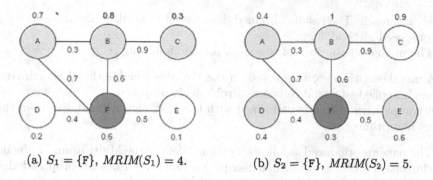

(a) $S_1 = \{F\}$, $MRIM(S_1) = 4$.   (b) $S_2 = \{F\}$, $MRIM(S_2) = 5$.

Fig. 2. Example of SN with $l = 1$ and 2 rounds where $MRIM(S_1 \cup S_2) = 6$.

A, B, and C. In this case, the available information in the adaptive model allows the method to increase the number of active nodes.

Figure 2 shows an example of an SN where 2 rounds are considered for the MRIM, with $l = 1$ and the Triggering Model. The value close to each node represents the value obtained in the Monte Carlo simulation to select the triggering set. In particular, the triggering set for a given node is defined as those adjacent nodes whose relation value is larger than or equal to the value obtained in the Monte Carlo simulation. Assuming that the Monte Carlo simulation value obtained for node F is 0.6, the triggering set for node F is $\{A, B\}$ since the value relation with A is 0.7, larger than 0.6, and the value of the relation with B is 0.6, which is equal to the value obtained in Monte Carlo simulation, 0.6.

The selected nodes are colored dark gray, while the influenced ones are colored light gray. In Fig. 2a node F is selected as the seed node, resulting in $S_1 = \{F\}$. Then, nodes A and B are directly influenced by F, and node C is also influenced by node B, resulting in an objective function value of 4. Notice that nodes D and E are not influenced by F since the relation value is smaller than the Monte Carlo value obtained for F (the value close to the node).

In the second round, depicted in Fig. 2b, the same seed node F is selected, $S_2 = \{F\}$. Notice that, in this round, the values obtained in the Monte Carlo simulation for each node (the value depicted close to each node) may vary. In this case, nodes A, B, D, and E are directly influenced, and node C is not influenced by B since the relation value is smaller than the Monte Carlo value obtained for B.

This work presents a novel metaheuristic approach to deal with MRIM, which allows us to find high-quality solutions in a reasonable computing time. Our main goal is to design an efficient algorithm to find the most influential users in an SN, considering the TM as an IDM, thus increasing the efficiency of the algorithm. An algorithm based on Variable Neighborhood Search (VNS) framework is presented, characterized by its efficiency when designing solutions for $\mathcal{NP}$-hard combinatorial optimization problems. The proposed procedure is validated over a set of real-world instances widely used in the context of social influence maximization and compared against the state-of-the-art method based on a totally greedy approach. The results obtained demonstrate the efficiency and efficacy of the proposed methodology.

The main contributions of this work are the following.

– A metaheuristic algorithm based on the Variable Neighborhood Search Approach applied to Social Networks Problems is proposed.
– A competitive testing is performed with the VNS algorithm and state-of-the-art algorithms.

The paper is organized as follows. Section 2 defines the Multi-Round Influence Maximization problem in this manuscript. Section 3 describes the proposed algorithm and the strategies used to solve it. Section 4 includes the computational results for both variants adaptive and non-adaptive. Finally, the conclusions and future research are discussed in Sect. 5.

2   Literature Review

Richardson [29] initially formulated the problem of selecting the target nodes in the SNs. Kempe et al. [12] were the first to solve the SNIMP formulating it as a discrete optimization problem. It has been shown that the SNIMP is $\mathcal{NP}$-hard [13]. Kempe et al. [12] proposed a greedy hill-climbing algorithm with an approximation of $1 - 1/e - \epsilon$, being e the base of the natural logarithm and ϵ any positive real number. This result indicates that the algorithm is able to find solutions which are always within a factor of at least 63% of the optimal value under the IDMs described in Sect. 1.

As a consequence of the computational effort required to evaluate the ICM, Kempe et al. [12] also proposed several greedy heuristics based on SN analysis metrics, such as degree and closeness centrality [32]. These methods only require one run of a Monte Carlo simulation to validate the single solution obtained using heuristic functions, thus increasing efficiency at the cost of loss of efficacy. When the metric considered is the degree of the node, the algorithm is called *high-degree heuristic*.

Several extensions of those first greedy algorithms were later proposed. In particular, Leskovec et al. [18] introduced the Cost-Effective Lazy Forward (*CELF*) selection that took advantage of the submodularity property to significantly reduce the run time of the greedy hill-climbing algorithm, becoming more than 700 times faster than the original procedure.

Goyal et al. [9] proposed a new algorithm called *CELF++* with the aim of improving the efficiency of the original CELF. It leans on the property of submodularity of the spread function for IDM, avoiding unnecessary computations. According to the authors, it is 35–55% faster than *CELF*.

Another related work is adaptive seeding [31], which uses the first-stage nodes and their accessible neighbors together as the seed set to maximize the influence, and is quite different from ours. In terms of the multi-round diffusion model and influence maximization, Lin et al. [19] focus on multiparty influence maximization where there must be at least two parties competing in networks.

The adaptive MRIM study follows the adaptive optimization framework defined by Golovin and Krause [8]. They also study adaptive influence maximization as an application, but the adaptation is at the per-node level: seeds are selected one by one. Later seeds can be selected based on the activation results of the earlier seeds, but earlier seeds would not help propagation again for later seeds. This makes it different from our multi-round model.

Finally in [33] and the extension [34] provide a further analysis about MRIM in an adaptive and non-adaptive way, compared with several methods in the state-of-the-art. The outperforming method it is a greedy method called AdaIMM that we used it to compare with our proposal.

Several surveys [1, 2] show that [34] is considered the state-of-the-art proposal, and metaheuristic approaches are scarce in the context of SNs problems.

3   Variable Neighborhood Search

Variable Neighborhood Search (VNS) is a metaheuristic [25] originally proposed as a general framework for solving hard optimization problems. The main contribution of this methodology is to consider several neighborhoods during the search and to perform systematic changes in the neighborhood structures. Although it was originally presented as a simple metaheuristic, VNS has drastically evolved, resulting in several extensions and variants: Basic VNS, Reduced VNS, Variable Neighborhood Descent, General VNS, Skewed VNS, Variable Neighborhood Decomposition Search, or Variable Formulation Search, among others. See [10, 11, 22] for a detailed analysis of each variant.

In this work, we propose a Basic Variable Neighborhood Search (BVNS) for providing high quality solutions for MRIM. This variant combines deterministic and random changes of neighborhood structures in order to find a balance between diversification and intensification, as presented in Algorithm 1.

Algorithm 1. $BVNS(k_{\max}, R, l)$

1: $S \leftarrow \emptyset$
2: **while** $R > 0$ **do**
3:     $S \leftarrow Construct(S, l)$
4:     $S \leftarrow LocalSearch(S)$
5:     $k \leftarrow 1$
6:     **while** $k \leq k_{\max}$ **do**
7:         $S' \leftarrow Shake(S, k)$
8:         $S'' \leftarrow LocalSearch(S')$
9:         $k \leftarrow NeighborhoodChange(S, S'', k)$
10:     **end while**
11:     $R \leftarrow R - 1$
12: **end while**
13: **return** S

The algorithm receives three input parameters: the largest neighborhood to be explored, k_{max}; the number of rounds used, R; and, finally, the size of the seed set per round l, resulting in the complete seed set of size $l * R$. The algorithm starts by creating the seed set S where it contains the selected nodes (step 1). Steps 2–12 represent the number of rounds used in the MRIM problem, and in step 3, an initial solution S is generated considering the constructive procedure presented in Sect. 3.1. The solution is then locally improved with the local search method described in Sect. 3.2 (step 4). Starting from the first predefined neighborhood (step 5), BVNS iterates until it reaches the maximum considered neighborhood k_{max} (steps 6–10). For each iteration, the incumbent solution is perturbed by the shake method (step 7). This method is designed to escape from local optima by randomly exchanging the position of k nodes, generating a solution S' in the neighborhood under exploration. The local search method (step 8) is then responsible for finding a local optimum S'' in the current neighborhood with respect to the perturbed solution S'. Finally, the neighborhood change method selects the next neighborhood to be explored (step 9). In particular, if S'' outperforms S in terms of the objective function value, then it is updated (i.e., $S \leftarrow S''$), and the search starts again from the first neighborhood (i.e., $k \leftarrow 1$). Otherwise, the search continues in the next neighborhood (i.e., $k \leftarrow k + 1$). The current round stops when reaching the largest neighborhood considered k_{max}, then the number of rounds it decreases in step 11 and starts the new round using the best seed set S. Finally, when the number of rounds ends, the best seed set found during the search is in S and returns this value (step 13).

3.1   Constructive Algorithm

Every VNS variant requires an initial solution, as stated by [11]. The original proposal of VNS [25] states that the initial solution does not affect to the quality of the obtained results, so it can be generated at random. However, several studies [20,28] have shown that providing a high-quality starting point helps the metaheuristic focus the search on more promising regions of the search space.

This work proposes a greedy algorithm which uses the objective function as greedy function value. This method calculates the number of activated nodes when selecting the next seed node u, $MRIM(u)$. With the aim of reducing the computational effort, the number of iterations required by Monte Carlo is reduced.

The nodes conforming the seed set for the first round are the ones that results in the maximum number of influenced users considering the reduced number of Monte Carlo iterations. Then, the method for selecting the seed sets for the next rounds depends on the approximation considered: the non-adaptive model or the adaptive model. In the former, the method selects the next l nodes most influential. On the contrary, in the latter, the nodes selected are those which are able to influence a larger number of non-previously influenced nodes.

3.2 Local Search

The improvement phase it is devoted to reach a local (ideally global) optimum. Designing a local search for influence maximization problems usually results in rather complex algorithms. In order to propose an efficient method, the local search proposed in this research is based on the one originally presented in [21], which allows us to have a short time-consuming local search procedure by limiting the search space explored.

The neighborhood of a solution S is defined as the set of solutions that can be reached by performing a single move over S. The move considered is a swap move $Swap(S, u, v)$ where node u is removed from the seed set, being replaced by v, with $u \in S$ and $v \notin S$. This swap move is formally defined as:

$$Swap(S, u, v) = (S \setminus \{u\}) \cup \{v\}$$

Thus, the neighborhood $N_s(S)$ of a given solution S consists of the set of solutions that can be reached from S by performing a single swap move. More formally,

$$N_s(S) = \{Swap(S, u, v) \quad \forall u \in S, \forall v \in V \setminus S\}$$

The size of the resulting neighborhood, $l \cdot (n - l)$, makes the complete exploration of the neighborhood not suitable for MRIM, even considering an efficient implementation of objective function evaluation. Then, the intelligent neighborhood exploration strategy proposed in [21] is followed, with the aim of reducing the number of solutions explored within each neighborhood. This reduction in the size of the search space is performed by exploring just a small fraction δ of the available nodes for the swap move.

The δ-value limits the number of nodes considered in the local search approach so it is recommended to select the most promising ones to be involved in the swap moves. In the context of MRIM, a node with a large out-degree can eventually influence a large number of nodes. Following this idea, we sort the candidate nodes to be included in the seed set in descending order with respect to their out-degree, while the candidate nodes to be removed from the seed set are sorted in ascending order with respect to their out-degree.

Notice that the objective function evaluation consists in a Monte Carlo simulation, being the most computationally demanding part of the proposed algorithm. For this reason, the proposed local search aims to limit the number of required simulations, thus leading to a more efficient procedure.

3.3 Shake

The perturbation mechanism in VNS is usually called the Shake procedure. The goal of this method is to diversify the search by generating a neighbor solution, which will not be explored by the local search method, and may eventually lead to further regions of the search space. We propose a method that modifies the structure of the solution according to a parameter k. Its value ranges from 1 to k_{max}, which is an input parameter of the complete procedure (see Algorithm 1).

The proposed shake method performs k swaps moves to the incumbent solution. As is customary in the BVNS methodology, these elements are selected at random. Recent works have studied more advanced shake techniques that balance diversification and intensification of the search. This strategy has been referred to as intensified shake (see [7] and [30] for further details).

4   Computational Results

This section describes the computational experiments designed to evaluate the performance of the proposed algorithms and analyze the results obtained. All experiments have been performed in an Intel Core i7-9750H (2.6 GHz) with 16 GB RAM and the algorithms were implemented using Java 17 and the *Metaheuristic Optimization framewoRK* (MORK) 13 [24]. The testbed of instances used in this work is the same set considered in the previous work. The results compare the performance of our proposal with the state-of-the-art greedy method AdaIMM [33].

Instances used by the state-of-the-art algorithm consist of two different datasets. On the one hand, the Flixster dataset is a network of the American social movie discovery service (www.flixster.com) where each user is represented by a node, and a directed edge from node u to v is formed if v rates a movie shortly after u does so on the same movie. The dataset is analyzed in [3], and the influence probability are learned by the topic-aware model. This instance contains 95969 nodes and 484865 directed edges. On the other hand, the NetHEPT dataset [5] is widely used in many influence maximization studies. It is an academic collaboration network from the "High Energy Physics Theory" section of arXiv from 1991 to 2003, where nodes represent the authors and each edge represents one paper co-authored by two nodes. There are 15233 nodes and 32235 directed edges in the NetHEPT dataset. All experiments are reproduced in the same environment thanks to the public code https://github.com/lichao-sun/Multi-Round-Influence-Maximization.

The Basic Variable Neighborhood Search (BVNS) parameters have been experimentally set, resulting in $l = 10$, $R = 5$, $\delta = 25$, $k = 0.1$. For both datasets, 100 Monte Carlo simulations are considered, as in the best previous method. Tables 1 and 2 contain the following performance metrics per round: the average objective function value (i.e., the number of nodes influenced, on average, after 100 simulations), Avg.; the average deviation with respect to the best known solution, Dev. (%); the average execution time of the algorithm measured in seconds, Time (s); and, finally, the number of times that the algorithm is able to reach the best solution in the experiment (#Best).

Table 1 shows competitive results when comparing both approaches in the non-adaptive version. In terms of deviation, AdaIMM reports 14.15% versus 2.37% of the BVNS. Notice that BVNS outperforms AdaIMM considering the Flixster instances in the average objective function. However, when dealing with NetHEPT instances, the average objective function value is more competitive. Analyzing the computational time, AdaIMM requires shorter computing time

Table 1. Results of the BVNS algorithm versus the state-of-the-art procedure in the non-adaptive version. Best results are highlighted with bold font.

Instance	R	BVNS				AdaIMM			
		Avg.	Dev. (%)	Time (s)	#Best	Avg.	Dev. (%)	Time (s)	#Best
NetHEPT	1	**355.89**	0.00	0.34	1	302.40	15.03	0.21	0
	2	**584.66**	0.00	0.49	1	544.40	6.89	0.28	0
	3	719.36	5.57	0.61	0	**761.80**	0.00	0.33	1
	4	900.89	6.72	0.75	0	**965.80**	0.00	0.51	1
	5	1015.53	11.41	0.82	0	**1146.30**	0.00	0.60	1
Flixster	1	**16637.78**	0.00	78.21	1	13560.51	18.50	6.59	0
	2	**17964.82**	0.00	89.16	1	13349.29	25.69	7.88	0
	3	**18288.07**	0.00	98.57	1	13655.16	25.33	9.50	0
	4	**18603.10**	0.00	106.82	1	14033.71	24.56	13.33	0
	5	**19120.03**	0.00	114.90	1	14820.86	22.49	17.22	0
Summary		**9419.01**	**2.37**	49.07	**7**	7312.96	14.15	**5.64**	3

(5.64 s versus 49.07 s), mainly due to the large number of nodes of Flixster instances, since AdaIMM only requires one complete Monte Carlo execution. Analyzing the number of best solutions found, BVNS is able to reach 7 out of 10 best results, while AdaIMM reaches 3 best solutions.

Table 2. Results of the BVNS algorithm versus the state-of-the-art procedure in the adaptive version. Best results are highlighted with bold font.

Instance	R	BVNS				AdaIMM			
		Avg.	Dev. (%)	Time (s)	#Best	Avg.	Dev. (%)	Time (s)	#Best
NetHEPT	1	**355.89**	0.00	0.34	1	302.40	15.03%	0.21	0
	2	**584.66**	0.00	0.52	1	557.20	4.70%	0.35	0
	3	**806.20**	0.00	0.68	1	776.90	3.63%	0.38	0
	4	913.99	6.93	0.81	0	**982.00**	0.00%	0.41	1
	5	1021.09	12.24	0.99	0	**1163.50**	0.00%	0.56	1
Flixster	1	**16637.78**	0.00	78.21	1	13611.37	18.19%	7.11	0
	2	**17831.39**	0.00	95.31	1	13749.83	22.89	7.99	0
	3	**18364.20**	0.00	106.87	1	13855.42	24.55	9.21	0
	4	**18929.40**	0.00	109.22	1	14213.69	24.91	13.24	0
	5	**19310.31**	0.00	121.23	1	14863.01	23.03	16.15	0
Summary		**9475.49**	**1.92**	51.42	**8**	7407.53	13.69	**5.56**	2

The results of the adaptive version, depicted in Table 2, are similar to the non-adaptive ones. The average deviation in AdaIMM is 13.69%, while BVNS obtains a value of 1.92%. Again, BVNS is more computationally demanding

(5.56 s versus 51.42 s). Analyzing the number of best solutions found, BVNS is able to reach 8 out of 10, while AdaIMM obtains 2 best solutions.

The results obtained in both non-adaptive and adaptive approaches show that BVNS is a competitive method for the MRIM when compared with the best method found in the literature, obtaining better solutions than AdaIMM. The main drawback of the proposal is the required computational time, although it can be limited by reducing the number of nodes explored in the local search method.

5   Conclusions

In this paper a Basic Variable Neighborhood Search algorithm for solving the adaptive and non-adaptive MRIM is presented. A constructive procedure based on the objective function is used with a surrogate local search based on swap moves. Since an exhaustive exploration of the search space is not suitable for this problem, an intelligent neighborhood exploration strategy is used which limits the region of the search space to be explored, focusing on the most promising areas. This rationale leads us to provide high quality solutions in reasonable computing time, even for the largest instances derived from real-world SNs commonly considered in the SNI problems. Since the intelligent neighborhood exploration strategy is parameterized, if computing time is not a relevant factor, the region explored can be easily extended to find better solutions, thus increasing the required computational effort. This fact makes the proposed MRIM algorithm highly scalable.

In our future work, we plan to increase the number of instances to provide robust results, reduce the computational time, study other methodologies as Variable Neighborhood Descent and General Variable Neighbourhood Search metaheuristic, and adapt the techniques developed in this work to influence minimization problems. This adaptation can be useful for minimizing the impact of fake news and monitor those users which can eventually transmit misinformation through the network.

Acknowledgments. The authors acknowledge support from the Spanish Ministry of Ciencia, Innovación y Universidades under grant ref. PID2021-125709OA-C22 and PID2021-126605NB-I00, Comunidad de Madrid and Fondos Estructurales of the European Union with grant references S2018/TCS-4566, Y2018/EMT-5062.

References

1. Aghaee, Z., Ghasemi, M.M., Beni, H.A., Bouyer, A., Fatemi, A.: A survey on meta-heuristic algorithms for the influence maximization problem in the social networks. Computing **103**(11), 2437–2477 (2021). https://doi.org/10.1007/s00607-021-00945-7
2. Banerjee, S., Jenamani, M., Pratihar, D.K.: A survey on influence maximization in a social network. Knowl. Inf. Syst. **62**(9), 3417–3455 (2020). https://doi.org/10.1007/s10115-020-01461-4

3. Barbieri, N., Bonchi, F., Manco, G.: Topic-aware social influence propagation models. In: 2012 IEEE 12th International Conference on Data Mining. pp. 81–90 (2012). https://doi.org/10.1109/ICDM.2012.122
4. Berger, J.: Word of mouth and interpersonal communication: A review and directions for future research. J. Consum. Psychol. 24(4), 586–607 (2014). https://doi.org/10.1016/j.jcps.2014.05.002
5. Chen, N.: On the approximability of influence in social networks. In: Proceedings of the Nineteenth Annual ACM-SIAM Symposium on Discrete Algorithms. p. 1029–1037. SODA '08, Society for Industrial and Applied Mathematics, USA (2008), https://dl.acm.org/doi/10.5555/1347082.1347195
6. D'angelo, A., Agarwal, A., Jin, K.X., Juan, Y.F., Klots, L., Moskalyuk, O., Wong, Y.: Targeting advertisements in a social network (Mar 2009), uS Patent App. 12/195,321
7. Duarte, A., Pantrigo, J.J., Pardo, E.G., Mladenović, N.: Multi-objective variable neighborhood search: an application to combinatorial optimization problems. J. Global Optim. 63(3), 515–536 (2014). https://doi.org/10.1007/s10898-014-0213-z
8. Golovin, D., Krause, A.: Adaptive submodularity: Theory and applications in active learning and stochastic optimization. J. Artif. Intell. Res. 42 (2010). https://doi.org/10.48550/arXiv.1003.3967
9. Goyal, A., Lu, W., Lakshmanan, L.V.S.: CELF++: Optimizing the greedy algorithm for influence maximization in social networks. In: Proceedings of the 20th international conference companion on World wide web - WWW '11. ACM Press (2011). https://doi.org/10.1145/1963192.1963217
10. Hansen, P., Mladenović, N., Brimberg, J., Pérez, J.A.M.: Variable Neighborhood Search, pp. 61–86. Springer, Boston (2010). https://doi.org/10.1007/978-1-4419-1665-5_3
11. Hansen, P., Mladenović, N., Pérez, J.A.M.: Variable neighbourhood search: methods and applications. Annals of Operations Research 175(1), 367–407 (oct 2009). https://doi.org/10.1007/s10479-009-0657-6
12. Kempe, D., Kleinberg, J., Tardos, É.: Maximizing the spread of influence through a social network. In: Proceedings of the ninth ACM SIGKDD international conference on Knowledge discovery and data mining. pp. 137–146 (2003). https://doi.org/10.1145/956750.956769
13. Kempe, D., Kleinberg, J., Tardos, E.: Maximizing the spread of influence through a social network. Theory Comput. 11(1), 105–147 (2015). https://doi.org/10.4086/toc.2015.v011a004
14. Khalil, E., Dilkina, B., Song, L.: Cuttingedge: Influence minimization in networks. In: Proceedings of Workshop on Frontiers of Network Analysis: Methods, Models, and Applications at NIPS. pp. 1–13. Citeseer (2013)
15. King, S.F., Burgess, T.F.: Understanding success and failure in customer relationship management. Ind. Mark. Manage. 37(4), 421–431 (2008). https://doi.org/10.1016/j.indmarman.2007.02.005
16. Klovdahl, A.S.: Social networks and the spread of infectious diseases: The AIDS example. Social Science & Medicine 21(11), 1203–1216 (1985). https://doi.org/10.1016/0277-9536(85)90269-2
17. Lawyer, G.: Understanding the influence of all nodes in a network. Sci. Rep 5(1) (2015). https://doi.org/10.1038/srep08665
18. Leskovec, J., Krause, A., Guestrin, C., Faloutsos, C., VanBriesen, J., Glance, N.: Cost-effective outbreak detection in networks. In: Proceedings of the 13th ACM SIGKDD international conference on Knowledge discovery and data mining. pp. 420–429 (2007). https://doi.org/10.1145/1281192.1281239

19. Lin, S.C., Lin, S.D., Chen, M.S.: A learning-based framework to handle multi-round multi-party influence maximization on social networks. In: Proceedings of the 21th ACM SIGKDD International Conference on Knowledge Discovery and Data Mining. pp. 695–704 (2015). https://doi.org/10.1145/2783258.2783392
20. Lozano-Osorio, I., Martínez-Gavara, A., Martí, R., Duarte, A.: Max-min dispersion with capacity and cost for a practical location problem. Expert Syst. Appl. **200**, 116899 (2022). https://doi.org/10.1016/j.eswa.2022.116899
21. Lozano-Osorio, I., Sánchez-Oro, J., Duarte, A., Cordón, Ó.: A quick GRASP-based method for influence maximization in social networks. J. Ambient. Intell. Humaniz. Comput. (2021). https://doi.org/10.1007/s12652-021-03510-4
22. Lozano-Osorio, I., Sanchez-Oro, J., Rodriguez-Garcia, M.Á., Duarte, A.: Optimizing computer networks communication with the band collocation problem: A variable neighborhood search approach. Electronics **9**(11), 1860 (2020). https://doi.org/10.3390/electronics9111860
23. Luo, C., Cui, K., Zheng, X., Zeng, D.: Time critical disinformation influence minimization in online social networks. 2014 IEEE Joint Intelligence and Security Informatics Conference, pp. 68–74 (2014). https://doi.org/10.1109/JISIC.2014.20
24. bibitemch9mork Martín, R., Cavero, S., Lozano Osorio, I.: rmartinsanta/mork: v0.13 (2022). https://doi.org/10.5281/ZENODO.6671107
25. Mladenović, N., Hansen, P.: Variable neighborhood search. Comput. Oper. Res. **24**(11), 1097–1100 (1997). https://doi.org/10.1016/S0305-0548(97)00031-2
26. Nguyen Hung, T., Thai My, T., Dinh Thang, N.: Stop-and-stare: optimal sampling algorithms for viral marketing in billion-scale networks. In: Proceedings of the 2016 International Conference on Management of Data, pp. 695–710. SIGMOD 2016, Association for Computing Machinery, New York, NY, USA (2016). https://doi.org/10.1145/2882903.2915207
27. Pérez-Peló, S., Sánchez-Oro, J., Martín-Santamaría, R., Duarte, A.: On the analysis of the influence of the evaluation metric in community detection over social networks. Electronics **8**(1), 23 (2019). https://doi.org/10.3390/electronics8010023
28. Pérez-Peló, S., Sánchez-Oro, J., Gonzalez-Pardo, A., Duarte, A.: A fast variable neighborhood search approach for multi-objective community detection. Appl. Soft Comput. **112**, 107838 (2021). https://doi.org/10.1016/j.asoc.2021.107838
29. Richardson, M., Domingos, P.: Mining knowledge-sharing sites for viral marketing. In: Proceedings of the eighth ACM SIGKDD International Conference on Knowledge Discovery and Data Mining, pp. 61–70 (2002). https://doi.org/10.1145/775047.775057
30. Sánchez-Oro, J., Pantrigo, J.J., Duarte, A.: Combining intensification and diversification strategies in VNS. an application to the vertex separation problem. Computers & Operations Research 52, 209–219 (Dec 2014). https://doi.org/10.1016/j.cor.2013.11.008
31. Seeman, L., Singer, Y.: Adaptive seeding in social networks. In: 2013 IEEE 54th Annual Symposium on Foundations of Computer Science, pp. 459–468. IEEE (2013). https://doi.org/10.1109/focs.2013.56
32. Stanley, W., Katherine, F.: Social Network Analysis. Cambridge University Press (Nov 1994). https://doi.org/10.1017/cbo9780511815478
33. Sun, L., Huang, W., Yu, P.S., Chen, W.: Multi-round influence maximization. In: Proceedings of the 24th ACM SIGKDD International Conference on Knowledge Discovery & Data Mining, pp. 2249–2258 (2018). https://doi.org/10.1145/3219819.3220101
34. Sun, L., Huang, W., Yu, P.S., Chen, W.: Multi-round influence maximization (extended version). (2018). https://doi.org/10.48550/ARXIV.1802.04189

A VNS Based Heuristic for a 2D Open Dimension Problem

Layane Rodrigues de Souza Queiroz and Thiago Alves de Queiroz(✉)

Institute of Mathematics and Technology, Federal University of Catalão,
Catalão-Go 75704-020, Brazil
{layanequeiroz,taq}@ufcat.edu.br

Abstract. This paper is related to open-dimension problems in the area
of cutting and packing. The problem we are interested in considers a set
of irregularly shaped items and a two-dimensional (2D) bin in which one
side is open. The objective is to pack all items in the bin and, in the
case of a bin with one opened side, we also want to minimize the length
of such a side. A packing cannot have items overlapping each other and
items extrapolating the bin's dimensions. This problem appears in the
metal-mechanic, textile, leather, and other related industries to the cutting of irregular pieces. We propose a variable neighborhood search-based
heuristic for such a problem. A solution is coded as a vector of items that
gives the sequence in which items will be packed. Neighborhood structures based on swap and insertion movements are considered in the local
search phase, while the shaking phase contains a single neighborhood
structure based on swap movements. Numerical experiments on benchmark instances show that the heuristic is competitive compared to other
literature methods, obtaining equal or better solutions for 90.90% of the
instances.

Keywords: Irregular cutting problems · Open dimension problems ·
Variable neighborhood search

1 Introduction

The problem of packing small objects inside one or more large objects appears in
many real-world applications. In the logistics area, for example, we have pallets
and boxes to be loaded into containers, trucks, trains, ships, or airplanes. This
problem is computationally hard in many of its variants and so mathematical
programming models and heuristics have been proposed in the literature. It is
important to mention that, from a theoretical point of view, it is equivalent to
cutting problems, which in turn requires the cutting of large objects to produce
small ones. In the textile and metal-mechanic industry, for example, fabric rolls
and metal plates are cut to produce pieces of products. For an overview of
packing and cutting problems, we refer to the book in [20].

In this paper, we are interested in problems where small objects can have
irregular shapes. We look for the packing (cutting) of all small objects (hereafter

A. Sleptchenko et al. (Eds.): ICVNS 2022, LNCS 13863, pp. 125–136, 2023.
https://doi.org/10.1007/978-3-031-34500-5_10

called items) in a single large object (hereafter called a strip). This problem is known as the two-dimensional irregular strip packing problem [18]. As the strip is assumed to have a fixed width, the objective is to minimize its opened length by obtaining a feasible packing. Concerning irregularly shaped items, the guarantee of a feasible packing can be obtained with geometric tools, such as the raster method, the phi-functions, the direct trigonometry, and the no-fit polygons. For these tools, we refer to the tutorial in [3]. We use the no-fit raster, a combination of the raster method and the no-fit polygons [23].

In the literature on the two-dimensional irregular strip packing problem, we may find different contributions, from simple mathematical programming models to sophisticated heuristics. As this problem is NP-hard, most of the contributions are related to heuristics. In [1], sequences of items are packed with the bottom-left rule. In this rule, an item is translated to the bottom and then to the left in the strip. This rule is also used in [15], where a genetic algorithm generates the sequence of items; in [10], where a 2-exchange heuristic generates the sequence of items; in [11], where simulated annealing is used for generating the sequence of items; in [17], where a biased random key genetic algorithm generates the sequence of items.

Other contributions are related to a constraint programming model in [4], the integration of the cuckoo search with a guided local search in [7], and a tailored branch-and-cut algorithm, where a variable neighborhood search heuristic generates feasible solutions, in [22]. Integer linear programming models are proposed in [5,8,16,19,23]. The model in [19] uses clique constraints to detect infeasible packings. They improved most of the previous solutions presented in the literature.

We propose a Variable Neighborhood Search (VNS) for the two-dimensional irregular strip packing problem. The VNS's neighborhood structures are based on swap and insertion movements. The shaking phase consists of a single structure based on swap movements, while the local search consists of the variable neighborhood descent (VND). The VNS generates the sequence of items, while a function is used to transform the given sequence into a feasible packing. For that, items are positioned in the strip by combining the bottom-left and top-left placement rules. Results obtained with the VNS are compared with those in [19,22], with better solutions for 27.27% of the instances and equal solutions for 63.63% of the instances.

The remainder of this paper is organized as follows. In Sect. 2, we define the problem and the geometric tools used to guarantee feasible packings. In Sect. 3, we present the variable neighborhood search and how a sequence of items is transformed into a problem solution. In Sect. 4, we perform computational experiments on literature instances and compare the performance of the VNS with the literature. In Sect. 5, we give some conclusions and directions for future works.

2   Problem Definition

This paper is about the Two-Dimensional Irregular Strip Packing Problem (2ISP). We assume the strip is rectangular, while items are defined as (irregular) polygons without holes. Each item j has a set of vertices V_j, an area a_j, and a reference vertex p_j. We assume an item is positioned in the strip by its reference vertex, which in turn is defined as the vertex with the lowest y-coordinate and, in the case of ties, with the lowest x-coordinate. A solution is built on the Cartesian plane. The strip's lower-left coordinates are at $(0, 0)$ and its top-right coordinates are at (∞, W). We associate the opened length (∞) to the x-axis and the width W to the y-axis. The problem's objective is to minimize the opened length while packing all items in the strip.

We assume the strip is discrete and then defined by a grid of points [2]. The reference vertex of items is positioned on points of this grid. A feasible solution (packing) is obtained when all items are packed inside the strip (i.e., there is no part/area of any item extrapolating the strip's dimensions) and items do not overlap each other (i.e., there is no intersection between any two items when positioned on the grid). To guarantee these two conditions to obtain a feasible solution, we calculate the inner-fit raster of each item with the strip and the no-fit raster between any two items. Figure 1 shows an example of irregularly shaped items, a rectangular strip, and the no-fit raster between two items.

The inner-fit and no-fit rasters are calculated in a pre-processing step [23]. For the inner-fit raster, each item j is positioned by its reference vertex at the lowest-left position on the grid, touching the strip's borders where possible but not extrapolating the strip's dimensions. Then, this item is translated around the strip always touching the strip's borders. The inner-fit polygon that is generated is next discretized according to the strip's grid. Positions having value "1" mean that such an item cannot be positioned there since it does not respect the strip's dimensions. For the no-fit raster, we consider each pair of items i and j. Item i is fixed on the plane, while j is positioned in such a way that it touches i. Then, item j is translated (by its reference vertex) around and touching i. The no-fit polygon that is generated is next discretized according to the strip's grid. Positions having value "1" mean that item j cannot be positioned there since such items will overlap each other.

3   Proposed Heuristic

We develop a VNS heuristic to the 2ISP. This heuristic has been applied to solve many continuous and discrete optimization problems, obtaining very competitive results compared to other literature methods [6,13]. Differently from other literature contributions that applied VNSs to irregular cutting problems, we consider the shaking phase defined on only one neighborhood structure, while the local search phase is composed of three neighborhood structures. The idea is to prioritize the local search phase to obtain high-quality solutions. As this phase could require a large computational time, the VNS has the advantage of

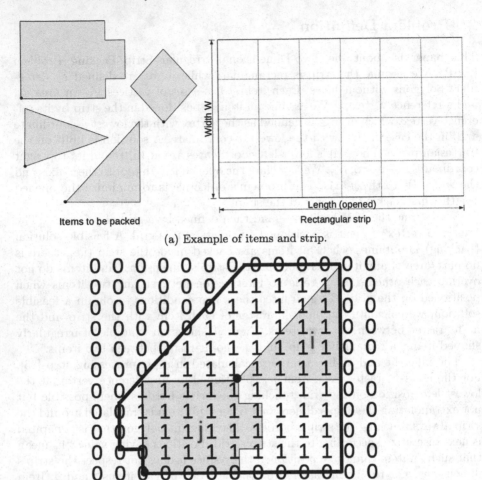

(a) Example of items and strip.

(b) Example of no-fit raster between items i and j.

Fig. 1. Illustrative example for the 2ISP.

carrying the optimization over a single solution and thus helping on reducing the computational effort. Algorithm 1 presents the proposed VNS.

In Algorithm 1, we code the solution x as a vector of integers. This is commonly adopted in the literature on irregular cutting problems [21,22]. Each integer represents the index of an item in the input instance. This means that x contains the sequence in which items are packed in the strip's grid. In the shaking phase, we consider that only the neighborhood structure N_1 is applied, where positions i and j are randomly chosen. On the other hand, the local-search phase considers three neighborhood structures, which are N_1, N_2, and N_3. In detail, the neighborhood structures are:

- N_1 (one-element swap): The elements of two given positions i and j are swapped, i.e., $i \leftrightarrow j$;
- N_2 (one-element insertion): Given two positions i and j, position i is inserted immediately after position j;
- N_3 (three-elements change): The elements of three given positions i, j, and u are changed, i.e., $i \rightarrow j$, $j \rightarrow u$, and $u \rightarrow i$. Notice that auxiliary variables are used to avoid losing information.

Algorithm 1: VNS PROPOSED TO THE 2ISP

---

1 $x \leftarrow$ randomly generated solution
2 **for** *a given number of iterations* **do**
3   **while** *true* **do**
        `/*shaking phase*/`
4          $x' \leftarrow$ random solution in the neighborhood structure $N_1(x)$
        `/*local-search phase*/`
5          $k \leftarrow 1$
6   **while** $k \leq 3$ **do**
7              $x'' \leftarrow$ first solution in the neighborhood structure $N_k(x')$ that is better than x', if one exists
8   **if** $F(x'') < F(x')$ **then**
9                  $x' \leftarrow x''$; $k \leftarrow 1$
10   **else**
11                 $k \leftarrow k + 1$
        `/*change of neighborhood*/`
12   **if** $F(x') < F(x)$ **then**
13             $x \leftarrow x'$
14   **else**
15   break

The local-search phase in Algorithm 1 consists of the variable neighborhood descent heuristic [12]. It starts by looking for the first solution in the neighborhood structure $N_k(x')$, initially for $k = 1$, that is better than x', the solution of the shaking phase. If this is true, solution x' is updated and the search continues on the same neighborhood structure; otherwise, the search continues on the next neighborhood structure. It is worth mentioning that all possibilities of positions in N_1, N_2, and N_3 are tested until finding the first improved solution if one exists. After the local search, solution x' is compared with the current solution x. The latter is updated if x' is better; otherwise, the while loop is broken.

The value of a solution x is determined by the function $F()$ in Algorithm 2. This function is based on the decoder proposed by [22]. The difference is that we are using only one placement rule, which is a combination of the bottom-left and top-left rules. The proposed placement rule divides the solution vector x into two parts. The left half of vector x assumes that items are positioned by the bottom-left rule, while the right half part has its items positioned by the top-left rules.

Algorithm 2: COST OF A SOLUTION x CALCULATED BY THE FUNCTION $F()$

---

1 **if** *solution x is in the hash table* **then**
2   ⌊ **return** *length L of the packing defined by x*

3 $S \leftarrow \emptyset$
4 **foreach** *item j in solution x* **do**
5   **if** *j is in the left half of x* **then**
6         $(a, b) \leftarrow$ the first point of the grid by the bottom-left rule that gives a feasible packing for j
7         $S \leftarrow S \cup \{j, (a, b)\}$
8   **else**
9         $(a, b) \leftarrow$ the first point of the grid by the top-left rule that gives a feasible packing for j
10   ⌊ $S \leftarrow S \cup \{j, (a, b)\}$

11 Save x and S in the hash table
12 **return** *length L of the packing S defined by x*

Figure 2 has an example of Algorithm 2 applied to a solution with four items in the sequence $\{1, 4, 3, 2\}$. Items 1 and 4 are in the left half of x and then are packed by the bottom-left rule. Items 3 and 2 are in the right half part and then are packed by the top-left rule. The resulting packing has length L, which is the cost of the solution returned by the function $F()$.

4   Computational Experiments

We coded all algorithms in the C++ programming language and performed computational experiments on literature instances. The experiments are executed in a computer with an Apple M2 processor, 8 GB of RAM, and macOS 13 as the operating system. The proposed VNS has a single parameter to define, which is the maximum number of iterations. We define it as a maximum time limit, set to 120 s, to solve each instance. The VNS runs 5 times and the best solution found among these is reported.

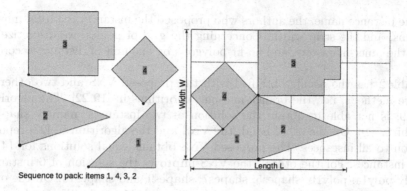

Sequence to pack: items 1, 4, 3, 2

Fig. 2. Example of packing obtained with the function $F()$ for a solution x.

Table 1. Data of the 22 instances.

Instance name	Authors	Number of items	Strip's width (W)
blazewicz1	Toledo et al. [23]	7	15
blazewicz2	Toledo et al. [23]	14	15
blazewicz3	Toledo et al. [23]	21	15
blazewicz4	Toledo et al. [23]	28	15
blazewicz5	Toledo et al. [23]	35	15
dagli1	Rodrigues and Toledo [19]	10	60
fu	Fujita et al. [9]	12	38
poly1a	Hopper [14]	15	40
poly1b	Rodrigues and Toledo [19]	15	40
poly1c	Rodrigues and Toledo [19]	15	40
poly1d	Rodrigues and Toledo [19]	15	40
poly1e	Rodrigues and Toledo [19]	15	40
shapes2	Toledo et al. [23]	8	40
shapes4	Toledo et al. [23]	16	40
shapes5	Toledo et al. [23]	20	40
shapes7	Toledo et al. [23]	28	40
shapes15	Toledo et al. [23]	43	40
shirts1-2	Rodrigues and Toledo [19]	13	40
shirts2-4	Rodrigues and Toledo [19]	26	40
shirts3-6	Rodrigues and Toledo [19]	39	40
shirts4-8	Rodrigues and Toledo [19]	52	40
shirts5-10	Rodrigues and Toledo [19]	65	40

We consider 22 instances from the literature, which may be found on the website of the EURO Special Interest Group on Cutting and Packing[1]. Table 1

---
[1] https://www.euro-online.org/websites/esicup/data-sets.

has the instance name, the authors who proposed the instance, the total number of items, and the strip width. Concerning the grid of points, we discretize the strip, the inner-fit rasters, and no-fit polygons by one unit of distance according to [19].

Table 2 has the results obtained with the proposed VNS and two other literature methods, i.e., the branch-and-cut algorithms in [19,22]. The algorithm in [19] is not able to obtain the solution of two instances, namely shapes15 and shirts5-10. On the other hand, the VNS and the algorithm in [22] report a solution to all instances. The proposed VNS obtains equal solutions for 14 out of 22 instances. For the others, the VNS improves the solution of 6 instances, namely poly1a, poly1b, shapes5, shapes7, shapes9, and shapes15. On the other hand, the VNS is worse for instances shirts2-4 and shirts3-6, differing from one unit in terms of length. The computing time of the VNS is not reported in the table because it is used as the stopping criterion and is equal to 120 s for each instance. In Fig. 4, we show the improved solutions obtained with the proposed VNS.

Table 2. Comparing the VNS with two other literature algorithms.

Instance	Proposed VNS	Souza Queiroz and Andretta [22]		Rodrigues and Toledo [19]	
	Length L	Length L	Time (s)	Length L	Time (s)
blazewicz1	8	8	7.43	8	0.01
blazewicz2	14	14	195.39	14	4.17
blazewicz3	21	21	3600.00	20	1139.96
blazewicz4	28	28	3600.00	27	3600.00
blazewicz5	35	35	3600.00	34	3600.00
dagli1	23	23	1.48	23	100.73
fu	34	34	3600.00	37	3600.00
poly1a	**16**	17	3600.00	17	3600.00
poly1b	**19**	20	3600.00	20	3600.00
poly1c	13	13	63.78	13	152.25
poly1d	13	13	3600.00	13	3600.00
poly1e	12	12	3600.00	12	3600.00
shapes2	14	14	2.10	14	1.09
shapes4	25	25	2431.50	25	3600.00
shapes5	**30**	31	3600.00	31	3600.00
shapes7	**41**	42	3600.00	45	3600.00
shapes9	**48**	49	3600.00	54	3600.00
shapes15	**61**	62	3600.00	–	3600.00
shirts1-2	13	13	0.03	13	0.02
shirts2-4	18	**17**	177.29	**17**	47.77
shirts3-6	25	**24**	3558.46	**24**	497.68
shirts4-8	33	33	3600.00	33	3600.00
shirts5-10	41	41	3600.00	–	3600.00

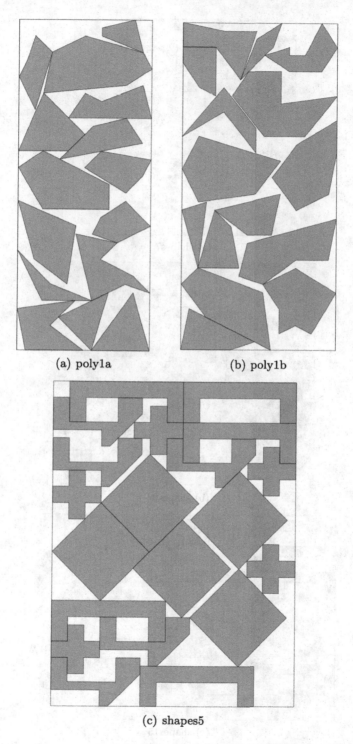

(a) poly1a   (b) poly1b

(c) shapes5

Fig. 3. Solutions improved by the proposed VNS - part 1.

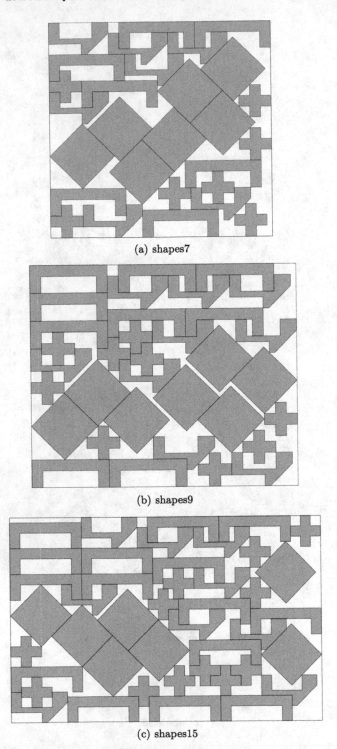

(a) shapes7

(b) shapes9

(c) shapes15

Fig. 4. Solutions improved by the proposed VNS - part 2.

5   Concluding Remarks

This paper is related to the two-dimensional irregular strip packing, an open-dimension problem, for which the strip's length is minimized while packing all items. Besides being an NP-hard problem, it is found in many real-world applications. As the items to pack may have an irregular shape, we use geometric tools such as the inner-fit raster and no-fit raster to guarantee feasible solutions. The proposed VNS has the shaking phase defined over a single neighborhood structure, while the local search as the VND has three neighborhoods based on swap and insertion movements. Due to the solution representation, we define a function to obtain the packing and so its length. In this function, items in the given sequence are packed by a combination of the bottom-left and top-left placement rules (Fig. 2).

The computational experiments on literature instances show the proposed VNS is competitive, obtaining equal or better solutions for 90.90% of the instances. For the other instances, the difference is one unit in the strip's length, which is relatively small. We notice that there is room for improvement in many directions. One could be in the proposal of new ways to code and decode a solution, as in the case of defining new placement rules. Further exploration of the scale adopted to the grid could also be worthwhile to identify the trade-off between solution quality and computing time. Another interesting direction is the combination of heuristics and mathematical programming models. It could be important to have a comparison between different paradigms, e.g., single trajectory heuristics versus population-based ones. In terms of instances, one direction could be to have items with holes and allow the rotation of items.

Acknowledgements. The authors acknowledge the financial support of the National Council for Scientific and Technological Development (CNPq grants numbers 405369/2021-2 and 311185/2020-7) and the State of Goiás Research Foundation (FAPEG).

References

1. Albano, A., Sapuppo, G.: Optimal allocation of two-dimensional irregular shapes using heuristic search methods. IEEE Trans. Syst. Man Cybern. **10**(5), 242–248 (1980)
2. de Almeida Cunha, J.G., de Lima, V.L., de Queiroz, T.A.: Grids for cutting and packing problems: a study in the 2D knapsack problem. 4OR **18**(3), 293–339 (2019). https://doi.org/10.1007/s10288-019-00419-9
3. Bennell, J.A., Oliveira, J.F.: A tutorial in irregular shape packing problems. J. Oper. Res. Soc. **60**, 93–105 (2009)
4. Carravilla, M.A., Ribeiro, C., Oliveira, J.F., Gomes, A.M.: Solving nesting problems with non-convex polygons by constraint logic programming. Int. Trans. Oper. Res. **10**, 651–663 (2003)
5. Cherri, L.H., Mundim, L.R., Andretta, M., Toledo, F.M., Oliveira, J.F., Carravilla, M.A.: Robust mixed-integer linear programming models for the irregular strip packing problem. Eur. J. Oper. Res. **253**(3), 570–583 (2016)

6. Duarte, A., Pardo, E.G.: Special issue on recent innovations in variable neighborhood search. J. Heuristics **26**, 335–338 (2020)
7. Elkeran, A.: A new approach for sheet nesting problem using guided cuckoo search and pairwise clustering. Eur. J. Oper. Res. **231**(3), 757–769 (2013)
8. Fischetti, M., Luzzi, I.: Mixed-integer programming models for nesting problems. J. Heuristics **15**, 201–226 (2009)
9. Fujita, K., Akagi, S., Hirokawa, N.: Hybrid approach for optimal nesting using a genetic algorithm and a local minimization algorithm. In: Proceedings of the 19th Annual ASME Design Automation Conference, pp. 477–484. Albuquerque, New Mexico, USA (1993)
10. Gomes, A.M., Oliveira, J.F.: A 2-exchange heuristic for nesting problems. Eur. J. Oper. Res. **141**(2), 359–370 (2002)
11. Gomes, A.M., Oliveira, J.F.: Solving irregular strip packing problems by hybridising simulated annealing and linear programming. Eur. J. Oper. Res. **171**(3), 811–829 (2006)
12. Hansen, P., Mladenović, N., Pérez, J.M.: Variable neighbourhood search: methods and applications. Ann. Oper. Res. **175**(1), 367–407 (2010)
13. Hansen, P., Mladenović, N., Todosijević, R., Hanafi, S.: Variable neighborhood search: basics and variants. EURO J. Comput. Optim. **5**, 423–454 (2017)
14. Hopper, E.: Mathematical models and heuristic methods for nesting problems. School of Engineering, University of Wales, Cardiff, Doutorado (2000)
15. Jakobs, S.: On genetic algorithms for the packing of polygons. Eur. J. Oper. Res. **88**(1), 165–181 (1996)
16. Leão, A.A.S., Toledo, F.M.B., Oliveira, J.F., Carravilla, M.A.: A semi-continuous mip model for the irregular strip packing problem. Int. J. Prod. Res. **54**(3), 712–721 (2016)
17. Mundim, L.R., Andretta, M., Queiroz, T.A.: A biased random key genetic algorithm for open dimension nesting problems using no-fit raster. Expert Syst. Appl. **81**, 358–371 (2017). https://doi.org/10.1016/j.eswa.2017.03.059
18. Oliveira, J.F.C., Ferreira, J.A.S.: Algorithms for nesting problems. In: Vidal, R.V.V. (eds.) Applied Simulated Annealing. Lecture Notes in Economics and Mathematical Systems, vol. 396, pp. 255–273. Springer, Berlin, Heidelberg (1993). https://doi.org/10.1007/978-3-642-46787-5_13
19. Rodrigues, M.O., Toledo, F.M.: A clique covering mip model for the irregular strip packing problem. Comput. Oper. Res. **87**, 221–234 (2017)
20. Scheithauer, G.: Introduction to Cutting and Packing Optimization: Problems, Modeling Approaches, Solution Methods, vol. 263. Springer, Cham (2017)
21. Souza, Queiroz, L.R., Andretta, M.: Two effective methods for the irregular knapsack problem. Appl. Soft Comput. **95**, 106485 (2020)
22. Souza, Queiroz, L.R., Andretta, M.: A branch-and-cut algorithm for the irregular strip packing problem with uncertain demands. Int. Trans. Oper. Res. **29**(6), 3486–3513 (2022)
23. Toledo, F.M.B., Carravilla, M.A., Ribeiro, C., Oliveira, J.F., Gomes, A.M.: The dotted-board model: a new mip model for nesting irregular shapes. Int. J. Prod. Econ. **145**(2), 478–487 (2013)

A Basic Variable Neighborhood Search Approach for the Bi-objective Multi-row Equal Facility Layout Problem

Nicolás R. Uribe[ID], Alberto Herrán[ID], and J. Manuel Colmenar[(✉)][ID]

Department of Computer Science and Statistics, Rey Juan Carlos University,
Mostoles, Spain
{nicolas.rodriguez,alberto.herran,josemanuel.colmenar}@urjc.es

Abstract. The Bi-Objective Multi-Row Facility Layout Problem is a problem belonging to the family of Facility Layout Problems. This problem is challenging for exact and metaheuristics approaches. We use the Pareto front approach instead of the weight approach by means of a non-dominated solution set which we update in order to keep only the non-dominated solutions. To tackle this problem, we propose a Basic VNS algorithm based on a constructive method that generates random solutions, a mono-objective local search that relies on an interchange move, and a shake method that applies insert moves. In this regard, we also explain how to adapt the mono-objective schema of the BVNS for a multi-objective one. Then, we compare our results with the state of the art and propose future work.

Keywords: Basic VNS · Bi-Objective optimization · Facility Location

1 Introduction

Facility Layout Problems (FLPs) are a well-known family of optimization problems with the goal of finding the optimal position of facilities in a given layout. See [1,9] and [10] for recent surveys. FLP is seen as a challenge for both exact and heuristic procedures. The very first work addressing this family of problems originally dates from 1969 [16] and was motivated by the need for a linear arrangement of different rooms along a corridor. This problem is called Single-Row Facility Layout Problem (SRFLP).

Many FLP variants can also be found in the literature. For example, in the Single-Row Equal Facility Layout Problem (SREFLP) [11] all the facilities have the same width. Using two rows, we find the Double-Row Facility Layout Problem (DRFLP) [4], and its variants. Other variants consider several rows (more than two rows) such as the Multi-Row Facility Layout Problem (MRFLP) [12],

---

This work has been partially supported by the Spanish Ministerio de Ciencia e Innovación (MCIN/AEI/10.13039/501100011033/FEDER, UE) under grant ref. PID2021-126605NB-I00.

and its variants. Another challenging variant is the Multi-Row Equal Facility Layout Problem (MREFLP) [2,18]. From the multi-objective point of view, the bi-objective MREFLP is a variant of the MREFLP which considers two objective functions and where both, the number of rows and the number of facilities that can be allocated in each row, are given by the instance. In other words, the layout configuration is fixed by the target instance.

In general multi-objective problems can be addressed from two different approaches. The first one combines all objective functions into a single one by means of a weighted sum, and returns a single value [3,6–8,13,15], and [17]. The second approach considers a set of different non-dominated solutions taking into account the objective functions separately [17]. In this work, we will follow this last approach.

The rest of the paper is structured as follows. In Sect. 2 we provide a description for the problem and an example of evaluation. Then, we describe our Basic VNS approach in Sect. 3. In Sect. 4 we explain our results and compare them with the state of the art. Finally, in Sect. 5, we present our conclusions and future work.

2   Problem Description

Given a set of facilities (F), a weight matrix corresponding to an objective function (W), a solution that we are going to evaluate (φ), the number of rows (m) and the number of columns (c), we can calculate the objective function value through Eq. 1 in this way: let $\rho(i,j)$ be the facility in row i and column j in φ, $w_{u,v}$ be the weight between facilities u and v in matrix W, and $d_{u,v}$ the distance between facilities u and v, the equation computes the sum of the products of all facilities pairwise weight and their pairwise distance.

$$\mathcal{F}(F,W,\varphi) = \sum_{i=1}^{m}\sum_{j=1}^{c}\sum_{k=1}^{m}\sum_{l=j+1}^{c} w_{\rho(i,j),\rho(k,l)} \cdot d_{\rho(i,j),\rho(k,l)} \tag{1}$$

As seen, the product has two parameters: a weight between the pairwise and the distance between these facilities. We extract the weight from the weight matrix, but we need to calculate the distance among them. In this problem, the Manhattan distance is used. This means that we will consider the vertical distance in addition to the horizontal distance between facilities. Manhattan distance is computed as shown in Equation (2).

$$d_{\rho(i,j),\rho(k,l)} = |l - j| + |k - i| \tag{2}$$

For this problem, we have two objective functions: *material handling cost, MHC* and *closeness ratio, CR*, and both are calculated with Equation (1). The difference between them is the weight matrix they use, which is different. Notice that these objectives are opposed due to these weight matrices values. Table 1 shows the weight matrix for *MHC* on the left and the weight matrix for *CR* on the right. Here we see that the weight between A and E facilities for *MHC* is

6, and it is -1 for CR. On the contrary, the weights between C and F are 1 and -1. So, depending on the values of these weight matrix, the objectives could be opposed.

Table 1. Weight matrices for the solution in Fig. 1. On the left, the weight matrix for the MHC. On the right, the weight matrix for the CR.

	A	B	C	D	E	F		A	B	C	D	E	F
A	0	5	3	2	6	4	A	0	4	6	2	-1	4
B	5	0	5	2	6	2	B	4	0	4	2	2	8
C	3	5	0	1	2	1	C	6	4	0	2	2	-1
D	2	2	1	0	2	2	D	2	2	2	0	6	2
E	6	6	2	2	0	6	E	-1	2	2	6	0	1
F	4	2	1	2	6	0	F	4	8	-1	2	1	0

For the sake of understanding we provide an example through Table 1 and Fig. 1. Let us begin with the contribution of the facility located in the first row and in the first column, $\rho(1,1) = $ A, to the objective functions. First, we need to calculate the distance between facilities A and D, where $\rho(1,2) = $ D. Notice that both facilities are in the same column, so, there is no vertical distance between them $(d_{AD} = |(1-1)| + |(2-1)| = 1)$. Then we proceed with the weight matrix. The contribution to the objective CR is $w_{AD} \cdot d_{AD}$, where $d_{AD} = 1$ and $w_{AD} = 2$. As a result, we have $w_{AD} \cdot d_{AD} = 2$. For the other objective function, we use the weight between A and D. Since $w_{AD} = 2$, we have the same result for the objective function MHC. We repeat this process for each pairwise between A and the other facilities. In the final step, we calculate the last pairwise facilities value, A and F. For these facilities, where $\rho(1,1) = $ A and $\rho(2,3) = $ F, the distance is $d_{AF} = |(3-1)| + |(2-1)| = 3$. The weight between facilities A and F is 4, for CR and for MHC. Then, the contribution of this pairwise for the objective function CR is $w_{AF} \cdot d_{AF} = 12$ and $w_{AF} \cdot d_{AF} = 12$ for the MHC one. In summary, to obtain the total contribution of facility A, we need the following operation:

$$w_{AD} \cdot d_{AD} + w_{AE} \cdot d_{AE} + w_{AB} \cdot d_{AB} + w_{AC} \cdot d_{AC} + w_{AF} \cdot d_{AF}$$

Once we finish calculating facility A, we have to repeat this process with the rest of facilities.

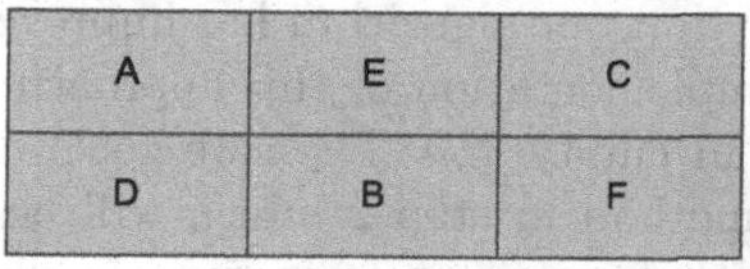

A	E	C
D	B	F

Fig. 1. Example layout with size 6.

3   Basic Variable Neighborhood Search

One of the most famous metaheuristics is Variable Neighborhood Search (VNS), which was proposed in 1997 by Mladenovic and Hansen [14]. This algorithm has several variants and, among them, one of the most used is the Basic Variable Neighborhood Search (BVNS). This variant lets us escape from the local optima through a *shake* movement. For mono-objective problems, we can apply the schema for BVNS as it was originally proposed. For multi-objective problems, we have to change this schema because this schema works for one solution, but not a set of solutions.

3.1   Bi-Objective BVNS

In this section, we will explain our algorithm proposal and the details of each proposed component for this algorithm.

If we work with mono-objective problems, it is simple to define if one solution is better than another. If we minimize, the solution with lowest value will be the best. If we are maximizing, then the other way around. For multi-objective problems, we have to compare the values for different objective functions. In this particular case, we have two functions that we have to minimize. One solution can dominate another, be dominated by another, or these solutions can be non-dominated among them. More precisely a solution φ_1 dominates φ_2 ($\varphi_1 \prec \varphi_2$) if the value of objective function $\mathcal{G}_i(\varphi_1)$ is better of equal for all objectives i, and exists one objective value where φ_1 is better. Equation (3) formally shows this concept.

$$\varphi_1 \prec \varphi_2 \; if$$
$$\forall i \in \{1..k\} : \mathcal{G}_i(\varphi_1) \leq \mathcal{G}_i(\varphi_2) \tag{3}$$
$$\land \; \exists i \in \{1..k\} : \mathcal{G}_i(\varphi_1) < \mathcal{G}_i(\varphi_2)$$

Due to the fact that we are going to work with more than one solution, we have to store them in a data structure. We will use a set of non-dominate solutions which we call *ND*. This set will contain only non-dominated solutions. Whenever we want to add a solution φ to this set, we will use the function *Update*, which will check if φ is dominated by any solution in the set. If so, we will not add this solution. If not, we will check all the solutions in the set, removing all the solutions dominated by φ.

One of the most important points of this algorithm is to define when we have improved *our solution*. To this aim, we will use the approach proposed [5]. This approach uses *ND* as the solution to be improved. If any changes have been made to *ND*, we consider that our algorithm has improved our current solution. Algorithm 1 shows our implementation of this approach.

In step 1) we create an empty *ND*. Then we generate a set of solutions by mean of a constructive method in step 2 which will be explained in Sect. 3.2. The number of solutions that we generate is given by the *maxCons* parameter. Then, in step 3, we populate *ND* with the solutions previously generated (S). It is worth noting that we *Update* each time we add a solution to our *ND*. Then, in

---

Algorithm 1: BVNS ALGORITHM (MAXCONS)

```
 1  ND ← ∅
 2  S ← Constructive(maxCons)                    ▷ Section 3.2
 3  ND ← Update(ND, S)
 4  S ← LocalSearch(ND)                          ▷ Section 3.3
 5  ND ← Update(ND, S)
 6  k ← 1
 7  while k < kMax do
 8  │   φ ← SelectRandom(ND)
 9  │   φ' ← Shake(φ, k)                          ▷ Section 3.4
10  │   S ← LocalSearch({φ'})                     ▷ Section 3.3
11  │   ND' ← Update(ND, S)
12  │   if (ND ≠ ND') then
13  │   │   k ← 1
14  │   └   ND ← ND';
15  │   else
16  │   └   k ← k + 1
17  return ND
```

---

step 4, we try to improve all our solutions as explained in Sect. 3.3. As a result, we obtain a new set S of improved solutions. Then, we update again our ND with S, obtaining the initial set of non-dominated solutions. In step 6, we set $k = 1$. Then, in step 7, we will iterate until we reach $kMax$. In this case, $kMax$ depends on the size of the instance. Inside the loop, in step 8, we select a random solution (φ) from ND. Then, in step 9, we shake our solution φ as described in Sect. 3.4. The parameter k indicates how many times we shake our solution. In step 10, we apply our local search procedure to φ', saving the resulting solutions in S. In step 11 we update our ND with S obtaining a new set ND'. In step 12 we check if ND' is equal to ND. If they are different then there was an improvement. Therefore, we update ND to ND' in step 11 and then, in step 13, we set $k = 1$ to reset the loop. If there was no improvement, k is incremented in step. Finally, in step 17, we return ND.

3.2   Constructive Method

In this first approach, we have used only one constructive method. With the aim of increasing diversity, our algorithm generates feasible solutions at random following a multi-start approach. This way, the non-dominated solution set will be populated with a number of different solutions. We have run this constructive 100 times, and we have stored these solutions in S. Once we have these solutions, we proceed to update our ND. Due to the fact that this set will contain only non-dominated solutions, ND could contain less than 100 solutions.

3.3 Local Search

Once we have a constructive method, we proceed to try to improve the solutions that we have obtained. We have implemented two different neighborhoods through two different moves: interchange and insert. In this section, we will explain the interchange movement. In the next section we will explain the insert one. Moreover, we will explain the local search that we have used. It is worth remembering that the layout is already defined by the instance, that means we know, as a constraint of the problem, the number of rows and the number of facilities per row.

Let us explain the interchange movement through Fig. 2. As stated in Sect. 1, the layout is already defined. Whenever we use a movement, we have to keep the same layout. On the left, we have the original layout, and on the right, we have the resulting one after the movement. The interchange movement changes the position of two facilities, leaving the others in their previous position. As we can see in Fig. 2, we have interchanged facilities D and C.

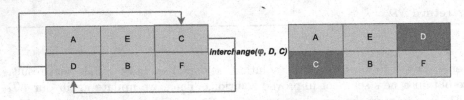

Fig. 2. Interchange facilities C and D.

For the local search, we will use the interchange movement. Our local search is based on the one proposed in [18] which considers only one objective. Due to the fact that we have two objectives, we need to adapt the implementation. The idea is to run two searches on each solution, one per objective. Let us explain in detail the local search through its implementation in Algorithm 2.

First, in step 1 we create an empty initial set of solutions S. Then in step 2, we will go through all the solutions in ND. We launch the local search that we mentioned above, focusing on the objective MHC and store this result in s_1. We do the same for the CR in step 4. Then, we add s_1 and s_2 to S in step 5 and iterate to the next solution in ND. Finally, we return S in step 6.

We represent this idea of the alternate local searches in Fig. 3. In this figure, we have two solutions, indicated as (b) and (a). In addition, we have also represented the two objectives MHC and CR. On the abscissa axis, MHC is represented; meanwhile, on the ordinate axis, CR is represented. There are four colored arrows. On the one hand, the orange ones represent how the solutions follow a trajectory toward the MHC objective. On the other hand, the blue ones follow a path toward the CR objective.

Algorithm 2: LOCALSEARCH(ND)

1 $S \leftarrow \emptyset$
2 **for** $i = 1$ *to* $|ND|$ **do**
3 $\quad$ $s_1 \leftarrow LS_{MHC}(ND[i])$
4 $\quad$ $s_2 \leftarrow LS_{CR}(ND[i])$
5 $\quad$ $S \leftarrow S \cup \{s_1\} \cup \{s_2\}$
6 **return** S

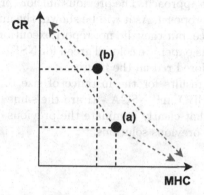

Fig. 3. Mono-objective local search.

3.4 Shake

In the previous section we have explained how the interchange movement works, and how we had used it in our local search procedure. In this section, we explain the insert movement. We have reserved this movement for the shake procedure. The reason is that this movement diversifies more than the interchange movement. Let us explain this movement through Fig. 4.

In Fig. 4 we have the original solution on the left, while on the right we have the one obtained after a certain insert. In particular, we have inserted facility A in the second row, third column. As we can see in this solution, all the other facilities but F have changed positions. The reason is that the layout has been previously defined and we have to keep this layout.

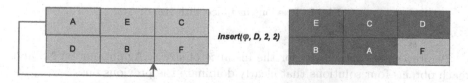

Fig. 4. Insert facility A in the second row and third column.

4   Results

In this section, we show our experimental results and then compare them with the state of the art. We represent our results graphically, in order to make the comparison easier for the reader.

The previous state of the art presented three algorithms: BBO, NSBBO and NSGA - II [17]. They used BBO as a weighted approach and NSBBO and NSGA - II as a Pareto front approach. In particular, NSBBO is the adaptation of the BBO to the Pareto front approach. The previous authors proposed four instances where we have run our proposal. As it will be shown, the authors reported only a few solutions per instance, but they did not report execution times. The solutions proposed by the previous paper are colored green for NSBBO and blue for NSGA - II. Our proposal is colored red en the figures.

Figure 5 shows the results for the instance of size 6. In this instance, the values reported for NSBBO and NSGA - II are the same. Our BVNS approach obtains four solutions that clearly dominate the previous ones. In fact, solution (71, 105) dominates all previous solutions.

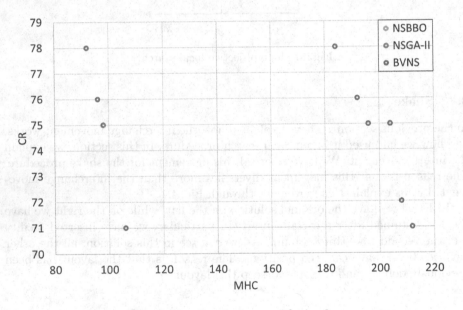

Fig. 5. Previous instance with size 6.

Figure 6 shows the results for the instance of size 8. Again, our BVNS approach obtains four solutions that clearly dominate the previous ones.

Figure 7 shows the results for the instance of size 12. In this instance, our BVNS proposal obtains eleven solutions that clearly dominate the previous ones.

Finally, Fig. 8 shows the results for the instance of size 15. In this instance, the values reported for the NSBBO and NSGA - II are more dispersed than in

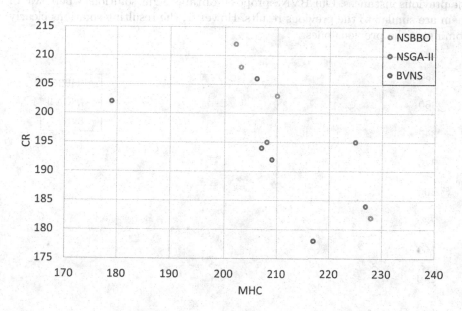

Fig. 6. Previous instance with size 8.

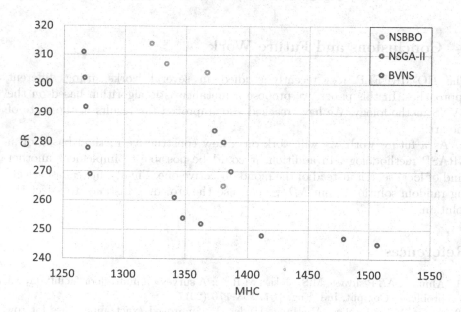

Fig. 7. Previous instance with size 12.

the previous instances. Our BVNS proposal obtains eight solutions where two of them are similar to the previous results. However, the resulting solutions clearly dominates the previous ones.

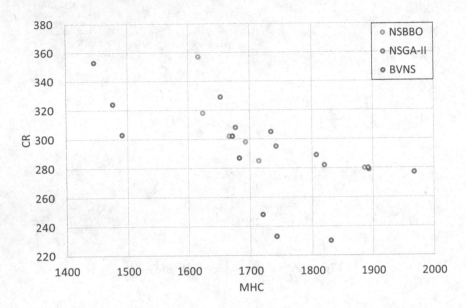

Fig. 8. Previous instance with size 15.

5   Conclusions and Future Work

The BO-MREFLP was recently studied in several works, using different approaches. In this paper, we propose a metaheuristic algorithm based on the BVNS methodology. We have reached and improved the results in the state of the art.

As a future work, we will work on a new constructive method based on a GRASP methodology. In addition, it could be possible to implement another kind of local search instead of the mono-objective one. Finally, instead of selecting random solutions from *ND*, we can use the crowding distance to select the solution.

References

1. Ahmadi, A., Pishvaee, M.S., Jokar, M.R.A.: A survey on multi-floor facility layout problems. Comput. Ind. Eng. **107**, 158–170 (2017)
2. Anjos, M.F., Fischer, A., Hungerländer, P.: Improved exact approaches for row layout problems with departments of equal length. Eur. J. Oper. Res. **270**(2), 514–529 (2018)

3. Chen, C.W.: A design approach to the multi-objective facility layout problem. Int. J. Prod. Res. **37**, 1175–1196 (1999)
4. Chung, J., Tanchoco, J.: The double row layout problem. Int. J. Prod. Res. **48**(3), 709–727 (2010)
5. Duarte, A., Pantrigo, J., Pardo, E.G., Mladenovic, N.: Multi-objective variable neighborhood search: an application to combinatorial optimization problems. J. Global Optim. **63**, 515–536 (2014)
6. Dutta, K.N., Sahu, S.: A multigoal heuristic for facilities design problems: Mughal. Int. J. Prod. Res. **20**, 147–154 (1982)
7. Fortenberry, J.C., Cox, J.F.: Multiple criteria approach to the facilities layout problem. Int. J. Prod. Res. **23**, 773–782 (1985)
8. Harmonosky, C.M., Tothero, G.K.: A multi-factor plant layout methodology. Int. J. Prod. Res. **30**, 1773–1789 (1992)
9. Hosseini-Nasab, H., Fereidouni, S., Fatemi Ghomi, S.M.T., Fakhrzad, M.B.: Classification of facility layout problems: a review study. Int. J. Adv. Manuf. Technol. **94**(1), 957–977 (2018)
10. Hosseini-Nasab, H., Fereidouni, S., Mohammad Taghi Fatemi Ghomi, S., Fakhrzad, M.: Classification of facility layout problems: a review study. Int. J. Adv. Manuf. Technol. **94**, 957–977 (2017)
11. Hungerländer, P.: Single-row equidistant facility layout as a special case of single-row facility layout. Int. J. Prod. Res. **52**(5), 1257–1268 (2014)
12. Hungerländer, P., Anjos, M.F.: A semidefinite optimization-based approach for global optimization of multi-row facility layout. Eur. J. Oper. Res. **245**(1), 46–61 (2015)
13. Matai, R., Singh, S., Mittal, M.: Modified simulated annealing based approach for multi objective facility layout problem. Int. J. Prod. Res. **51**, 4273–4288 (2013)
14. Mladenović, N., Hansen, P.: Variable neighborhood search. Comput. Oper. Res. **24**(11), 1097–1100 (1997)
15. Rosenblatt, M.J.: The facilities layout problem: a multi-goal approach. Int. J. Prod. Res. **17**, 323–332 (1979)
16. Simmons, D.M.: One-dimensional space allocation: an ordering algorithm. Oper. Res. **17**(5), 812–826 (1969)
17. Singh, D., Ingole, S.: Multi-objective facility layout problems using BBO, NSBBO and NSGA-II metaheuristic algorithms. Int. J. Ind. Eng. Comput. **10**(2), 239–262 (2019)
18. Uribe, N.R., Herrán, A., Colmenar, J.M., Duarte, A.: An improved GRASP method for the multiple row equal facility layout problem. Expert Syst. Appl. **182**, 115184 (2021)

Author Index

A
Abdelwanis, Moustafa 42
Aly, Ahmed 69

B
Benmansour, Rachid 14
Bolsi, Beatrice 97

C
Casado, Alejandra 1
Cavero, Sergio 82
Colmenar, J. Manuel 137

D
de Lima, Vinícius Loti 97
de Queiroz, Thiago Alves 97, 125
Duarte, Abraham 1, 27, 58, 112

G
Gabor, Adriana F. 69
González-Pardo, Antonio 27

H
Herrán, Alberto 137

I
Iori, Manuel 97

K
Kramer, Arthur 97

L
Lozano-Osorio, Isaac 112

M
Mladenović, Nenad 1, 42, 69

P
Pardo, Eduardo G. 58, 82
Pérez-Peló, Sergio 27

Q
Queiroz, Layane Rodrigues de Souza 125

R
R. Uribe, Nicolás 137
Robles, Marcos 82

S
Sánchez-Oro, Jesús 1, 27, 112
Sleptchenko, Andrei 42

Y
Yuste, Javier 58

Printed in the United States
by Baker & Taylor Publisher Services

Printed in the United States
by Baker & Taylor Publisher Services